# FIELDS OF GOLD

A Volume in the Series
*Cornell Series on Land: New Perspectives on Territory,
Development, and Environment*
Edited by Wendy Wolford, Nancy Lee Peluso, and Michael Goldman

A list of titles in this series is available at
www.cornellpress.cornell.edu

# FIELDS OF GOLD

## Financing the Global Land Rush

**Madeleine Fairbairn**

CORNELL UNIVERSITY PRESS    ITHACA AND LONDON

First published 2020 by Cornell University Press

Library of Congress Cataloging-in-Publication Data

Names: Fairbairn, Madeleine, author.
Title: Fields of gold : financing the global land rush / Madeleine Fairbairn.
Description: Ithaca [New York] : Cornell University Press, 2020. | Series:
    Cornell series on land : perspectives in territory, development, and environment |
    Includes bibliographical references and index.
Identifiers: LCCN 2019058742 (print) | LCCN 2019058743 (ebook) |
    ISBN 9781501750076 (hardcover) | ISBN 9781501750083 (paperback) |
    ISBN 9781501750090 (epub) | ISBN 9781501750106 (pdf)
Subjects: LCSH: Agriculture—Finance. | Agriculture—Capital investments. |
    Agriculture—Economic aspects. | Farms—Foreign ownership. | Financialization.
Classification: LCC HD1437 .F35 2020 (print) | LCC HD1437 (ebook) |
    DDC 333.33/5—dc23
LC record available at https://lccn.loc.gov/2019058742
LC ebook record available at https://lccn.loc.gov/2019058743

*For Hannah and Neil Fairbairn*

# Contents

# Acknowledgments

Although I cannot name them here, I would like to begin by thanking the many farmland investors, government officials, and activists who granted me interviews for this research. I am also deeply grateful to the investment conference organizers who allowed me to attend their events. This research would not have been possible without their generosity with time and information. I would also like to thank the many people who gave me permission to use their data and images in this book, including Jackie Tighe, Oliver Williams, Brandon Wu, and Bruce Sherrick, as well as representatives of Fquare, GMO, Savills Research, SLC Agrícola, NCREIF, and Preqin.

This research was also made possible by a number of grants and fellowships: a Graduate Research Fellowship from the National Science Foundation; an International Dissertation Research Fellowship from the Social Science Research Council; a Dissertation Completion Fellowship from the American Council of Learned Societies and the Andrew W. Mellon Foundation; a Louis and Elsa Thomsen Wisconsin Distinguished Graduate Fellowship from the College of Agricultural and Life Sciences at the University of Wisconsin–Madison; an Evelyn T. Crowe Fund Pre-dissertation Research Support Grant from the Community and Environmental Sociology Department at UW–Madison; a Scott Kloeck-Jenson Predissertation Travel Grant from the Global Studies Program at UW–Madison; and a Nave Short Term Field Research Grant from the Latin America, Caribbean, and Iberian Studies Program at UW–Madison. I greatly appreciate this generous financial assistance.

This book grew out of research conducted during my time in the Community and Environmental Sociology Department at the University of Wisconsin–Madison. There I benefited from the mentorship of Jack Kloppenburg and Gay Seidman. Jack is a role model for what a public intellectual can be—a brilliant thinker and writer whose ambitions extend well beyond thinking and writing. His unshakable and frequently expressed faith in my abilities was a marvelous gift. Gay's penetrating intellect, thoughtful feedback, and friendship were likewise indispensable. I wish that I could still workshop my articles over tea and cookies at her house. I am also deeply grateful to Jane Collins and Jess Gilbert, who both provided me with ideas that would become absolutely central to my thinking. I am also fortunate to have had Phil McMichael as an inspiring and supportive presence in my intellectual life since my days as an undergraduate at Cornell University and continuing throughout my graduate studies.

My research in Brazil was made possible by Bastiaan Reydon. In addition to sharing his broad expertise on Brazilian land markets, he housed, fed, and entertained me, enlivening my time in Brazil with his generous spirit and humor. In Brazil I would also like to thank Ana Karina Silva Bueno for friendship and hospitality, Vitor Bukvar Fernandes for research assistance, John Wilkinson for thoughtful feedback on my work, and Sérgio Sauer for generous advice while I was conducting research in Brasília. Before heading to Brazil I spent time researching the land rush in Mozambique, which helped shape my perspective. This research was kindly facilitated by David Stanfield of the Terra Institute and Calisto Ribeiro and Lourenço Duvane of the Rural Association for Mutual Support (ORAM).

Many others have supported me intellectually over the years in which this book took shape. Brett Christophers, Loka Ashwood, Keli Benko, and Mike Levien kindly read and provided feedback on earlier versions of this manuscript. Mike did so, heroically, over the winter holidays while working frantically toward a deadline of his own. Katy-Anne Legun and Zenia Kish, my tireless writing group, have read more versions of this than any of us can count. Many others have provided advice or intellectual stimulation along the way: Oriol Mirosa, Leland Glenna, Ryan Isakson, Oane Visser, Stefan Ouma, Geoff Lawrence, Jun Borras, and Kathryn De Master, to name a few. I am, of course, very grateful to the academic editors of the Cornell University Press Series on Land—Nancy Peluso, Wendy Wolford, and especially Michael Goldman—for agreeing to publish my work and to Jim Lance and Jennifer Savran Kelly for shepherding the book through the publication process. Two anonymous reviewers also provided me with tremendously thoughtful and constructive feedback.

I am also deeply indebted to my colleagues in the Environmental Studies Department at the University of California, Santa Cruz, including (but by no means limited to) Stacy Philpott, Carol Shennan, Deborah Letourneau, Jeff Bury, Sikina Jinnah, Daniel Press, and Andy Szasz. I have also greatly benefited from working with Emily Reisman, Halie Kampman, Allyson Makuch, Monika Egerer, Robin Lovell, Estelí Jimenez-Soto, Aysha Peterson, and the other brilliant graduate students of the UCSC agri-food working group. At UCSC I am particularly grateful for the friendship and inspiration provided by Julie Guthman and Margaret FitzSimmons, both of whom kindly read portions of this manuscript. Their intellectual fingerprints are all over my work—or at least the good parts of it. In a manner typical of her generous spirit, Margaret worked hard when she retired to ensure that her remaining research funds would be transferred to my name. It was this gift that allowed me to make the digital version of this book open access.

Finally, I wouldn't have completed (or even begun) this project without my family. I am grateful to my parents, Hannah and Neil, whose own intellectual curiosity never seems to diminish; my sister Catharine, who supplies ample laughter and commiseration as we forge our parallel academic journeys; and my babies Arthur and Edith, who have never known life when I wasn't writing this book and yet seem to love me anyway. Most of all I am grateful to my husband, Robbie. Doreen Massey and Alejandrina Catalano end the acknowledgments to their 1978 book *Capital and Land* with a tongue-in-cheek sentence: "Neither of us has a 'wife' to thank for solace and self-effacement." This commentary on the gender relations behind academic production, though undoubtedly still relevant, is happily inapplicable in my case. Thank you, Robbie, for the child rearing, the housecleaning, the hot meals, the graphic design help, the solace, and the self-effacement. I really am going to stop working on weekends soon.

# Abbreviations

| | |
|---|---|
| AAIMC | American Agricultural Investment Management Company Inc. |
| ABA | Agrifirma Brasil Agropecuária Ltda. |
| CFS | Committee on World Food Security of the FAO |
| CLT | community land trust |
| CNA | Agriculture and Livestock Confederation |
| FAO | Food and Agriculture Organization |
| FOE US | Friends of the Earth, US |
| GDP | gross domestic product |
| GMO | Grantham, Mayo, and van Otterloo |
| ha | hectares |
| HAIG | Hancock Agricultural Investment Group |
| INCRA | National Institute for Colonization and Agrarian Reform |
| IPO | initial public offering |
| IRR | internal rate of return |
| MEI | Indefinite Exploitation Module |
| MIGA | Multilateral Investment Guarantee Agency |
| MPT | modern portfolio theory |
| MST | Landless Workers' Movement |
| NCREIF | National Council of Real Estate Investment Fiduciaries |
| OCERS | Orange County Employees Retirement System |
| OTC | over-the-counter |
| PGIM | Prudential Global Investment Management |
| PRAI | Principles for Responsible Agricultural Investment that Respects Rights, Livelihoods, and Resources |
| PRI | Principles for Responsible Investment |
| PT | Workers Party |
| rai | Principles for Responsible Agricultural Investment in Agriculture and Food Systems |
| REIC | real estate investment cooperative |
| REIT | real estate investment trust |
| TIAA | Teachers Insurance and Annuity Association |
| TIMO | timberland investment management organization |
| VGGTs | Voluntary Guidelines on the Responsible Governance of Tenure of Land, Fisheries and Forests in the Context of National Food Security |

# Portuguese Terms

| | |
|---|---|
| *bancada ruralista* | rural caucus that represents large landowners in Brazilian Congress |
| *cartório* | public registry office |
| *cerrado* | Brazilian tropical savanna |
| *estrangeirização* | foreignization |
| *grilagem* | land grabbing via document fraud, often accompanied by violence |
| *jeitinho brasileiro* | the "Brazilian way" of getting around difficulties |
| *laranja* | front man, figurehead |
| *latifúndio* | large plantation |
| *quilombo* | communities descended from escaped slaves |
| *sesmaria* | Portuguese colonial land grant |

# FIELDS OF GOLD

# INTRODUCTION

**Buy land. They ain't making any more of the stuff.**

—Will Rogers (1879–1935)

On the third floor of a stately hotel, investment conference participants were spilling into a buzzing reception area. Long tables draped in white tablecloths held clusters of gleaming silver coffee urns surrounded by a lavish array of refreshments: tropical fruit salad, pastries, giant chocolate chip cookies, tiny crustless sandwiches, the works. The conference attendees—mostly white men between the ages of thirty and seventy, wearing suits in every conceivable shade of the same three colors—chatted in clusters against the walls and around linen-covered cocktail tables, some stalling in conversation before even making it to the coffee. At investment conferences like this, any conversation could lead to a lucrative investment deal or a new business partnership, so the coffee breaks are not really breaks at all. For half an hour, fund managers, corporate executives, and investors rubbed elbows to the sound of teaspoons tinkling against china until eventually a bell rang to announce the coffee break over.

The attendees gradually trickled through swinging doors into a large ball-room bordered by two tiers of gilded balconies and lit by an enormous crystal chandelier. There they seated themselves at tables facing the stage, where the next speaker was already being introduced. Thus far, the scene probably resembles every investment conference ever, but there was one major difference. The subject being discussed among all this finery was not the future of international banking or the latest in high-frequency trading. It was *farming*. These well-heeled men were in the market for dirt. The presenter now walking onstage was about to regale them with the particular benefits of buying farms in Ukraine, Australia,

1

Brazil, or the American Midwest. Others at the conference would discuss precision agricultural technology and irrigation systems. Farmland had somehow become an enticing new frontier for capital markets.

In recent years, the financial sector has developed a surprising interest in farms. Institutional investors—pension funds, university endowments, private foundations, and other organizations that manage huge pools of capital—are increasingly incorporating farmland into their investment portfolios. The same is true of those extremely wealthy people who in financial circles are euphemistically termed "high-net-worth individuals." This investor interest has spawned a host of new asset management companies eager to accommodate and encourage investors' newfound passion for soil. Promoting shiny new investment vehicles including farmland-focused private equity funds and real estate investment trusts (REITs), these managers promise to shepherd investor capital safely, and often extremely profitably, into plots of farmland the world over. This book examines why and how this transformation is taking place, drawing on several years of research on the global farmland investment industry, with a particular focus on two countries: the United States, which is a source of much investment capital and an established farmland investment target, and Brazil, which is an alluring, more frontier destination for international farmland investors.

In a process often referred to as "financialization," the financial sector has been deregulated, its profits have swelled, and it has gained unprecedented influence over nonfinancial companies. At the same time, nonfinancial companies are themselves increasingly being guided by financial logics and seeking out sources of financial return. I argue that we are now witnessing a "financialization of farmland," in which farms are being targeted for finance-sector investment and increasingly valued for their ability to produce financial profits. I trace the historical roots of this process and expose the institutions and discourses that make it possible.

I also argue, however, that farmland does not lend itself easily to becoming the next big financial asset class; the farmland investment industry must contend with moral sanctions surrounding landownership, with the inconvenient material attributes of its investment object, and with nationalistic policies regarding territorial sovereignty. Industry efforts to circumvent these obstacles, as well as the unintended consequences they produce, reveal that land's incorporation into global circuits of finance capital remains contingent and constrained. Farmland may be *treated* like a financial asset class, but it is still very far from becoming one. Still, incipient though the trend may be, growing financial-sector interest in farmland demands our attention; with so many livelihoods and identities dependent upon land, its incorporation into financial portfolios will have effects that reverberate through rural communities worldwide.

# The Global Land Rush

The investment conference scene described above captures just one dimension of a multifaceted rush for land that began slowly building steam around 2006 or 2007. This land rush first burst into the global spotlight in November 2008, when it was reported that the government of Madagascar had promised 1.3 million hectares (ha) of farmland—over half the arable land in the country, by some estimates—to the South Korean company Daewoo Logistics. The land, reports stated, was granted for a ninety-nine-year lease period and would be used to grow maize and palm oil for export to South Korea. That the government of this food-insecure country would grant such a large slice of its natural resource base to help feed and fuel a richer nation provoked concern among the international development community and outrage among the Malagasy people. In fact, this proposed land concession may have contributed to the success of a coup several months later; in one of his first official acts after taking office, the new president canceled the deal.[1]

Though this was an extreme case, it quickly became clear that it was not an isolated one. Large-scale land deals were occurring all over the world, a trend that activists, media, and scholars alike quickly dubbed a global "land grab."[2] These land deals were highly varied. The actors acquiring land included national governments and state-owned enterprises concerned about food security, as well as private corporations, investment firms, and wealthy individuals motivated solely by profit. The land was acquired through a variety of legal mechanisms; in countries with private property markets, land can be purchased outright, whereas in many African countries, land is owned by the state and must therefore be transferred via very long-term leases.[3] Ethiopia alone agreed to lease 2.5 million ha—an area almost the size of Belgium—to foreign investors, including 14,000 ha leased to the Saudi government–backed agribusiness Saudi Star for use in cultivating rice and 100,000 ha to Karuturi Global, an Indian company that proposed to grow cut flowers for the international market.[4]

Africa was a major target for land acquisitions, but so were many other parts of the globe. Investment funds and farming companies flocked to the highly fertile Black Earth region of Russia and Ukraine, known for growing wheat and sunflowers. Indonesia, Laos, and other Southeast Asian countries saw expanding oil palm and rubber plantations, while Brazil, Argentina, Paraguay, and other South American countries drew investments in large-scale production of soy, sugarcane, and more.[5] The amount of land changing hands is very difficult to calculate; the initial flurry of corporate press releases and newspaper articles about planned land deals included many that were ultimately canceled (like Daewoo's deal in Madagascar) or rapidly failed.[6] As of 2016, the Land Matrix—an independent monitoring initiative—reported that successfully completed land deals encompassed 57 million ha, while deals involving an additional 20 million ha were planned.[7]

The land rush was catalyzed, in large part, by an international food crisis. Already in February 2007, thousands of Mexicans gathered to demand relief from steep increases in the cost of corn-based tortillas. By 2008, the food crisis was global. In Senegal, police beat and teargassed protesters who could no longer afford rice. In Egypt, resentment toward the government was fueled by rising food costs, and the military was set to baking bread in the hopes of appeasing public anger. Grain prices decreased somewhat throughout 2009 and 2010, only to peak once again in 2011 and remain elevated through 2014.[8] These high crop prices had many causes. A sharp rise in global oil prices greatly increased the costs of agricultural production—including the costs of fertilizer, for which oil is a major input—and of transporting agricultural goods after harvest. At the same time, an extended Australian drought and other weather-related shocks led to production shortfalls, particularly of wheat. Additionally, government policies encouraging the use of biofuels created a food-fuel nexus in which basic food crops such as corn and soybeans became substitutes for fossil fuels. This forged a link between food and energy markets globally and heightened demand for land to grow both.[9]

However, food was not the only factor involved. Demand for agricultural land was also driven by concerns about water scarcity. By outsourcing agricultural production, arid countries like the Gulf States can increase their consumption of "virtual water"—the water that goes into producing consumer goods such as agricultural crops[10]—while conserving their own scarce water resources. At the same time, commercial demand for other types of natural resources led to large-scale land acquisitions for mining projects and even "green grabs" of forested lands for use as conservation areas and carbon sinks.[11]

This rush for land and the natural resources it harbors has set off alarm bells within the international development community and among human-rights activists. In many parts of the Global South, the land rights of the poor receive only the flimsiest of legal protections, making them susceptible to abrogation when a foreign investor begins making attractive overtures to their political representatives. Where governments own the land base, they can easily allocate large tracts to investors, dispossessing the peasant or indigenous communities that have traditionally depended on that land for their food and livelihoods.[12] To many governments, displacing poor rural people from their customary homes seems a relatively small price to pay for the agricultural jobs and economic development investors promise. In countries with private landownership, meanwhile, deep-pocketed investors constitute a new competitor for small farmers already struggling to afford land. The concern among critics of this land rush is that adding a sudden demand for land among rich states and corporations into agrarian contexts already characterized by deep inequalities could lead to the further marginalization of many rural peoples.

The pace of the land rush has subsided, particularly since 2014, when grain prices dropped from their historic highs. Yet while the peak of the land boom seems to have passed, its effects linger; in addition to the dispossession and deforestation cited by critics, I will argue that it has given rise to new institutional arrangements and changed the way that whole groups of actors think about land—changes whose effects are yet to be seen.

## The Financial Sector Discovers Farmland

My research homes in on just one aspect of this resource race: a growing interest in farmland on the part of the financial sector. The giant US pension fund Teachers Insurance and Annuity Association (TIAA), one of the largest players in the emerging farmland investment sector, illustrates how recently financial interest in farmland has emerged—and how rapidly it has grown. In 2007, TIAA (then known as TIAA-CREF) suddenly began acquiring enormous tracts of farmland as part of the investment portfolio it manages on behalf of retired teachers and other professionals. With hundreds of billions in assets under management (at the time; its assets have now surpassed a trillion dollars), this investment behemoth had the capital and the human resources to acquire an enormous portfolio of agricultural land rapidly. By 2012, just five years later, it already controlled $2.8 billion worth of farmland in the United States, Australia, Brazil, and Eastern Europe, making it one of the largest farmland owners and managers in the world. This included more than four hundred individual farm properties totaling 600,000 acres, most of them leased out to tenant farmers and operating companies. In the same year, TIAA announced the closing of a new fund, TIAA-CREF Global Agriculture LLC, with $2 billion in capital from third-party investors. This meant that TIAA was now not only buying farms for its own portfolio but also acting as an asset manager for other institutional investors, including Sweden's Andra AP-fonden pension fund and the Canadian financial institution British Columbia Investment Management Corporation. In 2015 TIAA added a second fund, this time with $3 billion in capital under management. By the end of 2017, less than a decade after its first farmland purchase, TIAA—a firm created to manage the retirement accounts of teachers—had come to control over 1.9 million acres of farmland worldwide.[13]

Though TIAA has thrown itself into the world of farmland investment with unparalleled vigor, it is not lacking for company. The farmland rush piqued the interest of many finance-sector investors, and a nascent industry grew up to cater to their needs. Agriculture- and farmland-focused investment conferences, which had barely existed in the past, suddenly proliferated and began drawing big crowds. Agricultural investment consultants started churning out reports on how and where to buy farms. The financial press marveled at farmland purchases by celebrity investors including Warren Buffett, George Soros, and Lord Jacob Rothschild.[14] News articles appeared with headlines like "Corn Farms Are Hotter Than New York Lofts," "Hedge Funds Muck in Down on the Farm," "Hot Money Turns from Stocks to Farmland," and "Betting the Farm."[15]

Finance-sector investors were drawn to farmland partly for the same reasons as everybody else—crop prices are high, ergo it pays to own a farm—but they also had their own reasons. The year 2008 saw not only a global food crisis but the worst financial meltdown since the Great Depression. As stock prices plummeted and venerable financial institutions teetered on the brink of collapse, farmland, along with other "real assets," took on a new luster in the eyes of investors. Real assets include such physical investments as real estate, infrastructure, art, and precious metals—in short, assets whose value stems from their substance or functions. This is in contrast to "financial assets"—those intangible assets whose value derives from an underlying contractual claim, such as stocks, bonds, certificates of deposit, and futures contracts. During economic downturns, when the value of financial assets can go up in a puff of smoke, worried investors often buy real assets. Reflecting this motivation, the financial headlines quoted above were joined by others that employed a common metaphor—"The New Black Gold: U.S. Farmland," "More Precious Than Gold? Farmland Has Glowing Appeal," "Fields of Gold: Investors Discover Lucrative Haven in Britain's Farmland." Those promoting farmland investment, meanwhile, often claim that it is like "gold with yield" or "gold with a coupon."[16] In general—in good economic times and bad—farmland has a reputation as an antidote for economic uncertainty. Because farmland prices are not highly correlated with the prices of stocks and bonds but do tend to increase with inflation, investment in farmland is used to reduce overall portfolio risk and hedge against inflation. Like gold, farmland is seen as a safe haven for capital.[17]

Farmland is so good at storing wealth in part because, as a finite asset, it tends to appreciate over time. The amount of potential farmland is limited by the earth's surface, and so growing human demand for food (not to mention fiber and fuel) tends to put upward pressure on land prices. Land price increases can be particularly dramatic on urban outskirts and in places where economic growth is causing rapid urbanization.[18] My focus in this book is on a growing investor demand for farmland qua farmland, not for the purposes of urban development, yet—as I will argue—the allure of increasing property values is still by no means incidental to its appeal.

## Financialization and the Agri-food System

In recent decades, finance has increasingly become the gravitational center of economic life. In the process of financialization, financial actors and financial motives have taken on a greater role across economic sectors.[19] The rise of finance can be traced, at least in part, to the economic policy shifts associated with neoliberal restructuring, particularly the deregulation of the US financial sector.[20] Until the 1970s, finance was viewed as playing a supporting role to productive economic activity; banks, for instance, provided the important services of converting deposits into business loans, supplying short-term liquidity to business, and vetting the creditworthiness of loan applicants.[21] Government regulations sought to constrain finance, allowing it to facilitate the circulation of capital and therefore economic activity, but restricting financial concentration and risk taking. Beginning in the 1970s, however, many long-standing financial regulations were repealed. Limits on credit card and other interest rates were removed, as were restrictions on bank mergers and prohibitions on cross-industry activity that had formerly prevented banking, insurance, and investment activities from taking place within the same firm.[22] During this time, there was also an enormous increase in global capital mobility, beginning in 1971 with the end of the Bretton Woods system of fixed exchange rates and continuing throughout the 1980s and 1990s as the International Monetary Fund pressured developing countries to liberalize their capital markets.[23] Formerly constrained to a supporting role, with these changes finance increasingly became the main act.

At the same time, a major shift was taking place in the relationship between corporations and their investors. Beginning in the 1970s, "shareholder value" became the new shibboleth of corporate management; this principle dictates that a company's value is defined not by the number of people it employs or the amount it produces but by its ability to generate returns for its shareholders. Corporations, which had once been masters of their own domain, were now expected to prioritize making money for their investors above all else.[24] To ensure that they took this lesson to heart, corporate managers were increasingly compensated in stock options, aligning their financial interests with those of their shareholders. During the 1980s, companies that failed to deliver impressive returns found themselves vulnerable to a hostile takeover by "corporate raiders" from the financial sector. In the decades that followed, corporate buyouts were continued, albeit in friendlier form, by private equity companies, which use investor capital to acquire companies, restructure them, and resell them at a profit. Corporations were encouraged to be lean, achieving high returns while owning as few tangible assets as possible; any "underutilized assets" were seen as an invitation for private equity to take over the company and sell them off.[25] Critics of the shareholder value revolution argue that it has fostered a pervasive short-termism, with corporations sacrificing long-term investment in productive capacity in pursuit of impressive quarterly earnings reports and frequent payouts to shareholders.[26]

This period also saw a proliferation of increasingly creative and complex financial assets for investors to choose from. There has been, for instance, explosive growth in markets for financial derivatives—financial assets whose value is *derived* from fluctuations in the value of underlying assets, which could be commodities, interest rates, currencies, or any manner of other things. Another font of novel financial instruments is "securitization," a process in which the cash flow from a pool of underlying assets is aggregated into a single income stream that is then used as collateral in issuing bonds to investors.[27] Almost anything that produces income can be securitized, from student-loan payments to overdue parking fines to revenues from old Italian films, and the search for ever more unorthodox asset streams to securitize is one of the hallmarks of the financialization era.[28] The downside of all this financial ingenuity became evident in 2008, when the shaky financial edifice composed of mortgage-backed securities and their associated derivatives came tumbling down, triggering a financial crisis.[29]

Though the changes associated with financialization are diverse, they share a common theme: the growing centrality of *financial income* in the form of earned interest, dividends, and capital gains on investments relative to the *productive income* earned by actually making and trading commodities. In fact, Greta Krippner defines financialization as "the tendency for profit making in the economy to occur increasingly through financial channels rather than through productive activities."[30] The ascendancy of financial over productive income, she argues, is occurring on two fronts. First, the profits made by financial sector firms such as banks and asset management companies have been steadily growing as a proportion of the economy as a whole. Second, even nonfinancial firms that previously made their money from production or trade have become increasingly dependent on the portfolio income they generate from financial activities. Rather than just selling cars, auto companies increasingly make money by financing car loans; rather than just selling plane tickets, airlines increasingly make money by investing in energy derivatives.[31] Even regular individuals increasingly embrace the pursuit of financial income. Instead of putting money in a savings account and leaving it there untouched, middle-class households in the Global North are now active participants in their own wealth management, constantly fiddling with their stock portfolios and refinancing their homes, with their ability to take a vacation, or even retire, increasingly tethered to the vagaries of the stock market.[32]

The agri-food system, like every other economic sector, has increasingly been permeated by financial actors and restructured according to financial motives.[33] Beginning in the 1980s, regulators loosened long-standing limits on speculation in markets for agricultural commodity derivatives. These derivatives—which include forwards, futures, options, and swaps—are based on the price of globally traded agricultural products such as wheat, corn, and sugar.[34] The deregulation of agricultural commodity derivative markets allowed an influx of financial capital, which, critics argue, may have contributed to the food crisis of 2008 by increasing volatility in global food markets.[35] Meanwhile, finance has also been reshaping the food retail industry, where private equity companies have been purchasing supermarket chains—including Safeway, Albertsons, Winn-Dixie, and Bi-Lo in the US alone—and restructuring them to improve shareholder profits.[36]

As in other sectors of the economy, it is not only financial companies that are engaging in financial activities and prioritizing financial profits. Food retailers like Walmart, Tesco, Carrefour, and Kroger now offer a range of financial services, from credit cards to check cashing to money transfer services.[37] Grain traders, too, have branched into financial activities. The big grain-trading companies—the so-called ABCDs: Archer Daniels Midland, Bunge, Cargill, and Louis Dreyfus—have long used commodity derivative markets to hedge their business activities, but in recent years they have come to see derivative trading as a source of profit in itself, opening financial subsidiaries that offer asset management services to investors.[38] Financial activities, which formerly took place behind the scenes of food production and trade, are increasingly becoming the primary focus.

Until recently, however, the steady creep of finance capital appeared to stop at the farm gate. Investing in corn futures is one thing. Buying a corn farm is quite another. Indeed, the general modus operandi of finance in recent decades has been to *move away* from such messy entanglements with physical ownership.[39] Though finance can never be fully divorced from productive assets, the main thrust of financial innovation has been in securitizing and trading the income streams those assets produce.[40] Nonetheless, a few years into the twenty-first century big investors began buying farms.

# Unlikely Bedfellows: Financial Investment in Farmland

This book seeks to answer a number of questions about the financial sector's growing appetite for farmland. Why, over the last decade, has farmland become uniquely appealing to investors? What new institutional developments are making it possible for investors with little knowledge about land to find and acquire properties all over the world? What kinds of language and ideas are being used to legitimize farmland investment and attract investors? What role do national-level politics and policy play in mediating the relationship between global capital and domestic land markets?

This topic is, I believe, important for both intellectual and practical reasons. First, on the intellectual side, it presents a fascinating study in contrasts. It is hard to come up with two things more different than farmland and finance. Farmland is tangible and immobile, and its profitability stems directly from its materiality. It evokes images of hard work and seems timeless, perhaps even a little old-fashioned. Idioms like "to bring down to earth" and "to ground" equate land with all things sensible. Farms are a popular topic for picture books, their basic components so fundamental as to be considered suitable reading material for the preliterate. Finance, on the other hand, is highly mobile and seemingly abstract. As an economic sector it has a reputation for innovation and risk taking. Most people have only a rudimentary understanding of finance, and it becomes more recondite with each wave of financial innovation.

Aside from having little in common with esoteric investment products, farmland has always been an awkward commodity in its own right. Its physical characteristics—immobility, heterogeneity, and expansiveness—make it unlike any other good, while agriculture, with its seasonality, unpredictable weather, and perishable end product, is unlike any other production method.[41] Because land was not created for sale on the market, political economist Karl Polanyi famously labeled it a "fictitious" commodity and argued that the privatization of land tends to give rise to political backlash.[42] Any movement to free land markets from legal and normative constraints—to disembed them from social institutions—would, Polanyi believed, necessarily give rise to a counter-movement demanding their re-embedding via the creation of protective laws.[43] Land has enormous cultural significance, standing in different times and places as a link to a valued agrarian way of life, a spiritual home, and the resting places of ancestors, to name just a few.[44] Land also acts as territory—a crucial symbol of national sovereignty and pride.[45] This ungainly commodity will not tuck neatly into financial portfolios, and it is fascinating to watch the attempt in progress.

An emerging body of scholarship explores these efforts at making land conform to the needs of finance. Whereas the commodification of land described by Polanyi involved the privatization of common lands—a process often termed "enclosure"—farmland is now being transitioned to the more advanced stage of commodification required by investors. These efforts—variously described as "rendering land investible" by Tania Li, "asset making" by Oane Visser, and "assetization" by Antoine Ducastel and Ward Anseeuw—seek to translate farmland into a source of predictable, comprehensible, politically neutral future returns for investors.[46] Yet, their success is by no means assured. As the above authors, as well as Sarah Sippel, Stefan Ouma, André Magnan, and others observe, the farmland fund managers and other corporate executives working to turn farmland into a new financial asset class face a messy biophysical reality, as well as legal and social movement challenges, and questions of moral legitimacy. As Ouma puts it, the transformation of farmland into a financial asset class must be seen as an ongoing "practical accomplishment" of finance capital, one whose success is by no means assured.[47] Furthermore, the ways in which finance and farmland interact are geographically uneven, as varied as the historical and political contexts within which land is located.[48] Sifting the developing relationship between farmland and finance can tell us a lot about the institutional, discursive, and legal processes through which markets in nature are made.

There is, however, a second, much more important reason for researching the changing relationship between farmland and finance: its possible effects on land and wealth distribution. There are no exact data available on the amount of financial-sector capital so far invested in global farmland markets; a 2010 study commissioned by the OECD estimated it to be in the range of $10 billion to $25 billion, while a 2012 report by Macquarie Agricultural Funds Management already put the figure substantially higher, at $30 billion to $40 billion.[49] Though these figures may sound impressive, they are fairly paltry from the perspective of the financial sector. Consider that, in 2016, pension funds alone controlled over $36 *trillion* globally.[50] Or that the $9.5 billion of its own assets that TIAA had invested in farmland by the end of 2017 constituted under *1 percent* of the trillion dollars the company controls.[51] In this context, $40 billion is truly small potatoes. Or, as a hedge fund manager once explained to me: "It's mouse nuts." When this phrase drew a blank look from me, she added definitively, "It's spit."

But financial-sector spittle has the potential to buy a lot of land. In Iowa, where farmland cost an average of $7,943 per acre in 2014, $40 billion could buy slightly over 5 million acres—approximately *a sixth of the state's farmland*. In Brazil's soy frontier state of Piauí, where farmland cost around $1,000 an acre in 2014, it would buy well over *half the state*. In Mozambique, meanwhile, where the nationalized land base is not officially for sale but can nonetheless be purchased by various means, one estimate put farmland at around $200 an acre in 2012, meaning that $40 billion could buy an area the size of the *entire country*.[52]

In short, context matters when assessing the magnitude of capital involved; what looks like chump change to the financial sector may be anything but in poor rural areas. It is therefore important to interrogate growing financial interest in farmland now, while it is still a nascent phenomenon and not yet a major driver of land prices. At present, some intrepid investors are just dipping their toes into farmland markets to see how it feels, but if they like the experience, their massive friends may follow suit, heaving in their multibillion-dollar limbs as customary land users are displaced by rising land prices or government expropriation.[53]

Ultimately, the prospect of landownership concentration in the hands of the financial sector matters because it has the potential to propel economic inequality. Land, unlike a business empire, cannot be created through hard work, market savvy, or even good luck. It preexists human effort and can only be gradually amassed by those with the wherewithal to corner it. Yet, once owned, land is a source of free money for the owner; she can now charge rent to anyone wishing to use her land, and the land will become only more and more valuable with time as population grows and economic development proceeds all around. In other words, land serves as a way to generate income from wealth. The work of Thomas Piketty reveals that wealth plays a crucial role in perpetuating economic inequality. Since the late 1970s, Piketty demonstrates, the proportion of national incomes accruing to the holders of capital has been increasing relative to the proportion earned through labor. In other words, those who make money simply because they already have money are winning out over those who make money by working. This suggests that, unless governments do something to reverse the trend, wealth will continue to concentrate disproportionately in the hands of the rich.[54]

Land is important to this picture of growing economic inequality because property is one of the major forms in which wealth is stored, and because the rent and capital gains (i.e., the profits from appreciating land value) it produces are key sources of income from wealth.[55] But land is not *just* another way to produce income from wealth. The rich may treat land like gold, but it is not gold. For the non-wealthy, land is a crucially important source of food and livelihoods. It matters greatly, therefore, if the financial institutions that control much of the accumulated wealth of society decide that land is a preferred route for storing that wealth and generating income from it. In the Global South, their hefty demand for land could fuel dispossession by encouraging privatization of lands used by peasant or indigenous communities without official land rights. Financial demand for land may also contribute to more gradual and insidious forms of dispossession. Even in wealthy countries where private property in land is the norm—the US, for example—small farmers often survive on a knife's edge, struggling to pay their rent or meet their mortgage payments. This struggle is even more pronounced for indigenous, black, and other farmers of color, who must contend with the racism that permeates the institutions governing land access, on top of the day-to-day challenges of making a living as a farmer. Their dream of landownership will only move further out of reach if they find themselves in regular competition with hedge funds. It is for these reasons that it is important to study this topic, even if the amount of finance capital invested in farmland at present is "mouse nuts."

## Research Methods

Understanding the financial sector's increasing interest in farms requires "studying up"—making elite actors and worlds the study subject. In a seminal 1972 essay, anthropologist Laura Nader argued that her discipline was spending too much time studying poor people and not enough time examining the wealthy and well-connected—that it was time to turn the spotlight on "the colonizers rather than the colonized, the culture of power rather than the culture of the powerless, the culture of affluence rather than the culture of poverty."[56] Since then, anthropologists and other social scientists have set their sights on Wall Street investment banks, the Chicago Board of Trade, and more, producing powerful insights about the internal logics and cultures of finance.[57] Studying up means viewing finance as a realm of cultural production, not just business logic. It requires understanding that, just as peasant farmers do not have a purely holistic and spiritual relationship to land, financiers are not merely capitalist automatons driven by the need to maximize profit margins.

Studying financial investment in farmland is uniquely challenging because it involves a global trade in something that is entirely fixed in place. My research, which took place primarily between 2010 and 2014, sought to understand the financialization of farmland as a transnational set of practices and imaginaries that nonetheless take shape—are formed, modified, contested, and hybridized—within particular places.[58] They bear the markings of idiosyncratic, place-based histories and are confronted by local and national politics in the places they are implemented. To understand this "actually existing" farmland financialization, I combined research on the global farmland investment industry with national-level case studies of the US and Brazil.[59]

To discern the motivations and mechanisms behind the financial sector's newfound enthusiasm for farmland at a global level, I interviewed fifty-one people involved in the farmland investment industry.[60] They came from several groups. *Investors* control capital, and buying farmland is one of their many options for what to do with it. *Asset managers* create investment vehicles—some sort of investment fund or company—designed to attract investor capital and use it to buy farmland. Through the investment vehicle, they buy properties, develop them, and either lease them out to tenant farmers or contract with a property management company to farm the properties on their behalf. Alternatively, investors and asset managers may buy into farmland through agricultural *operating companies* that own farms. These agribusinesses may be privately held or publicly listed on the stock market. Finally, *intermediaries* such as agricultural investment consultants and conference organizers play an important supporting role, forging connections and shaping shared beliefs about farmland.[61]

To add to my understanding of the farmland investment industry, I attended agricultural investment conferences as a participant observer. These conferences attracted attendees from all over the world, revealing transnational dynamics that might have been hard to get from individual interviews alone. I attended panels and presentation sessions, dropped by informational booths, and had a great many conversations—what some would call "informal interviews"—during coffee and cocktail hours. Attending these events allowed me to absorb the latest thinking on farmland investment and to pinpoint the major actors and their investment strategies. The conferences also proved to be crucial sites for constructing meaning—where narratives about the global land rush take shape and the identity of the nascent sector is honed.[62] I also assiduously collected and read investment reports and other industry documents. Together, these methods allowed me access to the world of the private agricultural investor, comprehending an infant industry in its own terms. This industry-level research gave me a fairly solid picture of global developments, though one weighted toward Western agricultural investors and the Americas.[63]

In addition to studying why and how the financial sector is buying farmland internationally, I was interested in how this investment is shaped by the histories and politics of particular countries.[64] I chose to study the US as a window onto the historical development of the global farmland investment industry. The US was ground zero for many of the changes associated with financialization— the financial deregulations, the changing ideas about corporate governance, the investment-happy populace. It is a global center for financial activity and boasts an unparalleled concentration of investor capital; over half of global pension fund assets are managed by US institutions, while fully 65 percent of hedge fund assets globally are under US management.[65] As with all things finance, US institutions have been at the forefront of fostering finance-sector investment in farmland. Some of today's largest global farmland investors got their start buying American farms, and US agricultural land remains popular with institutional investors. At the same time, US-based financial institutions are among the biggest investors in farmland internationally. The financial analysis company Preqin reports that 38 percent of natural resource fund managers investing in agriculture or farmland were based in North America as of 2016, and that of the ten asset managers that raised the most capital for farmland and agriculture-focused funds between 2006 and 2016, six were US-based.[66]

My research on the US case was largely historical, drawing from an array of primary and secondary sources. These included newspaper articles, congressional hearing records, and land price data, as well as the work of historians. This research was also augmented by insights from my interviews with US-based farmland investors and asset managers (discussed above), some of which took place in New York City, where I lived for much of the time I was researching and writing this book, while some occurred in other locations or, when unavoidable, by phone.

I chose to study Brazil, meanwhile, as a window onto the present, globalized land rush. When I began thinking about this topic in 2009, Brazil was the poster child for international farmland investment, owing to its agro-export dominance, its vast land area and concentrated ownership structure (which together allow for the rapid acquisition of large tracts of land), and its free-market approach to agriculture and trade. Investors are particularly drawn to the *cerrado* region, an immense tropical savanna that stretches through the middle of the country and has, in recent decades, become the breadbasket of Brazil, blanketed with fields of soy, as well as cotton, corn, coffee, cattle, and more. As of 2016, the Land Matrix monitoring initiative listed Brazil as the fifth-biggest "target country" for large-scale land acquisitions by area, though my experience suggests that among finance-sector investors this ranking might be even higher.[67] Brazil is also a particularly interesting place to study the relationship between government and foreign investors. It has been a perennial case study for research on the "developmental state," particularly how the state can channel foreign investment capital in such a way as to contribute to sustained national development.[68] Its national debate over whether and how to restrict foreign land purchases therefore provides insight into the delicate regulatory challenge posed by transnational farmland investment.

I spent seven months in Brazil between 2010 and 2012, most of it split between Brazil's primary business hub, São Paulo, and its capital city, Brasília. In São Paulo, I interviewed operating company executives and farmland fund managers (discussed above), whereas in Brasília I conducted twenty-eight interviews with the policy makers, politicians, and activists working to shape Brazil's policy response to the global land rush.[69] While in Brazil I also conducted other types of fieldwork. I spent a day touring a Brazilian farm with representatives of foreign financial institutions, I attended congressional meetings on the regulation of foreign farmland investment, and I spent a week in Western Bahía State, where many new farmland investments are located.

Access is always difficult when studying up. Elite actors are shielded from academic scrutiny by security personnel, legal teams, and prohibitive costs of entry. The agricultural investment conferences I attended, for instance, had registration fees ranging from $375 to over $1,000 per day, which, of course, does not include all the associated travel and hotel expenses. I was able to be there only because generous conference organizers let me in at steep discounts—I twice paid 50 percent of the registration fee for a conference but generally was allowed in for free— or in exchange for services like helping with conference registration or being a "microphone runner" during question-and-answer sessions.[70] Once at these events I found it relatively easy to mingle. While some people would sidle away upon discovery that I was not a representative of the University of Wisconsin endowment looking to invest large sums of money, others were happy enough to take a break from networking to explain some element of investment strategy to a researcher.

I made initial contacts for interviews through networking at these investment events, as well as by cold e-mailing prominent companies. Additional interview contacts came from "network sampling"; at the end of each interview I would ask the participant for suggestions of other industry contacts. In some cases, this approach was incredibly effective, as with the Brazilian fund manager who sent introductory e-mails to three of his good college friends, each now at the helm of another major Brazilian operating company or farmland fund. Willingness to be interviewed varied by actor type. Investors were most reluctant to be interviewed, likely because they spend their days fending off overtures from those seeking their time and capital. It was easier to get asset managers and operating company executives to grant interviews, I expect because the need to raise capital puts them in the habit of accepting any opportunity to talk about their projects. In general, industry participants became increasingly leery of granting interviews as the project progressed, owing to growing negative media coverage of the land rush.

My positionality as a white, educated American was a major asset when it came to research access. The industry actors I interviewed were mostly white, male, middle-aged, and relatively wealthy. Many were undoubtedly multimillionaires. Yet despite this power differential, we had quite a lot of cultural overlap. I chatted with fellow conference attendees about life in New York City and was several times invited out to dinner with groups of them at the end of the day. My educational background was particularly useful; the fact that my two alma maters—Cornell and the University of Wisconsin—are both well-known agricultural schools gave me some credibility with the Brazilian operating company executives I interviewed, almost making up for my lamentable decision to study sociology. My identity as a young woman, meanwhile, played a more mixed role. In the male-dominated professions of finance and agriculture, I was a bit of an oddity, which may have increased people's willingness to talk to me. A woman interviewing men also works well with existing cultural norms for gendered interaction in both my study countries. As Sarah Babb puts it, "There is an established gender dynamic in conversation: the female role is to ask eager questions about a man's life, and the man, flattered by the focus of female attention, holds forth at great length." A woman interviewing men is therefore "following the steps of a well-understood cultural dance."[71] My male research participants might not have spoken at such length or shown such patience with my initial ignorance had I been male. The downside of being a woman interviewing men was that it heightened the already uneven power dynamics within interviews. I had participants grab my recorder to turn it on and off, interrupt to ask me questions about my personal life, and commandeer the interview direction with some regularity.[72]

I have taken various measures to protect my research participants. With the exception of elected officials, the people I interviewed are identified by pseudonym or general characteristics, and in a couple of cases I have even altered identifying features in my description of participants in order to give them plausible deniability about their involvement in the research. In doing participant observation I always identified myself in a transparent manner as a sociologist researching growing investor interest in farmland, and when anyone asked questions about my research I did my best to respond openly. Yet my research subjects hardly constitute the kind of vulnerable populations for whom such research protections were invented; the educated, technologically savvy people I interviewed have no difficulty in finding my publications and—as more than one interview participant casually reminded me—have the resources to sue me if they don't like what I write. As a result of this power imbalance, I am probably more circumspect in what I have written than was strictly necessary. I don't name the particular investment events I attended. I also use company names sparingly; they are never attached to the statements of their executives, and when writing about particular firms I rely on publicly available information whenever possible. Nonetheless, I have tried, in writing this book, to retain some critical distance, remembering that my accountability as a researcher also extends to the communities and natural spaces affected by the land rush.

## Book Organization

Farmland is many things to many people. Economically, it is *both* an essential means of production that can produce agricultural income *and* a scarce asset that generates rental income and passive appreciation. In other words, it is valued both for its ability to produce commodities and as a scarce commodity itself. At the same time, as a source of shelter, sustenance, and identity, land abounds with value that cannot be measured in dollars.

These multiple ways of valuing land coexist in a state of uneasy tension, and land booms are the perfect time to study the interactions and contradictions between them. Land booms advance the ongoing process of land commodification but also catalyze movements to reclaim land as something more than a commodity. They are transformative, leaving behind changed economic arrangements, changed legal structures, and changed landscapes. One of the most important legacies of the recent global land boom, I will argue, is a heightened awareness of agriculture and farm real estate among financial actors; the financial sector has developed new ways of thinking about, discussing, and investing in farmland that will shape how future land booms play out. The land rush also foreshadowed what may be a growing challenge for national governments: whether and how to regulate investor interest in farmland.

The book begins by considering farmland financialization from two angles: as the growing penetration of financial institutions into farmland markets (chapter 1) and as the unique mind-set that financial landowners bring to their farmland purchases (chapter 2). It then examines the obstacles that complicate this process: the *moral* associations of farmland (end of chapter 2), its *material* peculiarities (chapter 3), and the *political* tensions surrounding its ownership and use (chapter 4). Detailing investor efforts to overcome these constraints reveals farmland as highly contested terrain whose annexation into the world of finance is far from being a foregone conclusion. Along the way, the book toggles back and forth between the national and international levels. It begins with a history of farmland investment in the US, where many international investors got their start (chapter 1), then zooms out to explore motivations and strategies of the nascent global farmland investment industry (chapters 2 and 3), and finally zooms back in to examine national-level political resistance in Brazil (chapter 4).

Although the financial sector's interest in farmland appeared to come out of nowhere in 2008, it had in fact been building for some time, spurred on by prior land market booms and busts. To demonstrate this, chapter 1 explores the evolving relationship between private institutional investors and farmland markets in the US, where some of today's biggest global farmland investment firms originated. In the US, institutional investors have long profited from farmland, but they have generally done so indirectly, via farm mortgage lending. Land booms and busts were always moments of turmoil in which lenders might dabble in direct investment or wind up with a great number of foreclosed farms on their hands, but for the most part financial investors were content to leave landownership with the farmers. This began to change, however, in the aftermath of the devastating 1980s farm crisis, with the appearance of financial management companies specializing in farmland investment. By the time the simultaneous food and financial crises hit in 2007–2008, the groundwork for the financialization of farmland had been laid and there was an explosion of new farmland investment funds. With more investment capital at their disposal than they could easily deploy in the US, these farmland investors expanded their search internationally.

But the financialization of farmland extends beyond the simple creation of more agricultural investment vehicles; I argue in chapter 2 that it also involves particular understandings and depictions of the value of land. This chapter revisits classic and modern theories of land value to shed light on the complex interactions between land's productive and financial qualities that appeal to investors. While investors are attracted by the evident materiality and productivity of agricultural land, most nonetheless depend on the profits from land appreciation to make their investments worthwhile. In this sense, they treat land as a source of future capital gains, much like any other financial investment. At the same time, however, the pursuit of moral legitimacy leads them to emphasize the ways in which their work can be seen as productive activity.

Though investors are largely interested in farmland for its financial qualities, those qualities come wrapped in an extremely cumbersome package. Chapter 3 explores the various strategies that the farmland investment industry has so far devised to get around the awkward materiality and temporality of land and agriculture. These range from portfolio diversification strategies to new ways of trading farmland on the stock market. In seeking to free the profitability of farmland from its material moorings and reduce *agricultural* uncertainty, however, the financial community may inadvertently increase the *financial* uncertainty associated with farmland investment.

The free flow of investment capital into domestic farmland markets can also be complicated by national politics. To explore this, I follow investor capital into another of its most popular target countries: Brazil. When the land rush came to Brazil, it erupted into a controversy over the "foreignization" of rural land, culminating in a new government regulation restricting the amount of property that foreign investors could acquire. This controversy and regulatory shift is the subject of chapter 4. The government's restrictions on foreign investment, I argue, used concerns about national sovereignty, which are seen as the legitimate purview of the state, as a substitute for concerns about land speculation and the scale of agricultural production, whose legitimacy as regulatory objects is under constant attack. These regulations were singularly ineffective; no sooner were the new restrictions in place than foreign investors began finding ways to get around them. Brazil's experience highlights the enormous challenge that states face when it comes to regulating flows of finance capital into land markets.

The conclusion considers some existing approaches to curbing farmland financialization: corporate codes of conduct, activist campaigns targeting particular investors, and efforts to democratize investment through cooperative landownership. It argues that, to ensure the well-being of rural communities and of agro-ecosystems, farmland investment must be subject to democratic control. This means that the political discourse surrounding farmland investment must not be constrained to the dominant "politics of productivity," in which ownership is automatically considered justified if the land is put to use (and the more intensive use the better). Instead, what we need is a genuine and expansive public dialogue about the benefits and risks of farmland financialization.

# FARMLAND INVESTMENT COMES OF AGE

> And it came about that owners no longer worked on their farms.
> They farmed on paper; and they forgot the land, the smell, the feel
> of it, and remembered only that they owned it, remembered only
> what they gained and lost by it.
>
> —John Steinbeck, *The Grapes of Wrath*, 1939

While most of the investment conferences I attended were of the tablecloth-and-chandelier variety, in 2011 I went to an unassuming regional farmland investment event in the US Midwest. Here jeans and plaid replaced suits and ties, and those promoting farmland investment were more likely to tout opportunities in the wilds of Wisconsin than in sub-Saharan Africa. Meandering through the exposition hall, I met Ingrid, a family farmer turned personal financial consultant to farm families. When I explained that I was studying growing investor interest in farmland, Ingrid shook her head and said that the current atmosphere reminded her of the 1970s. Back then, she said, bankers were pushing midwestern farmers to expand their operations by taking out additional farm loans and mortgages. But then interest rates shot up to 21 percent, grain prices fell, and farmland values plummeted. When farmers could no longer pay their debts, the bankers called them fools and took their land. "A lot of people in the market now don't remember the late seventies well," she concluded.

Ingrid's comments serve as a reminder that the current wave of farmland investment can only be understood within historical context.[1] In the US, private financial institutions have a long history of involvement with agriculture, though the nature of that involvement has changed over time.[2] Historically, institutional investors were involved in farmland markets primarily as farm mortgage providers. Rather than buying farms themselves, they used their extensive capital to facilitate land purchases by farmers—a crucial distinction. Yet the farmland investment industry did not materialize out of nowhere in

2008 either. In the US, it has been building slowly since the 1980s when those same farm mortgages led many farmers into foreclosure and dispossession. Its rise has paralleled shifting ideas about the role of finance capital in the economy as a whole—from supplying capital to producers to making the greatest possible profit for shareholders. These gradual developments prepared the ground so that, when conditions were ripe for another land boom, the farmland investment sector could rapidly flourish.

I focus in this chapter on the historical development of the US farmland investment industry. Because many of the biggest global players got their start investing in US farmland, and because US farmland is still a highly desirable target for investment, this history sheds considerable light on the drivers behind the global land rush.[3] It reveals a gradual shift in how the financial sector has approached farmland—from collateral on producer loans to investment object in its own right—and explores the reasons that the search for land went global. It also shows that the recent land rush cannot be understood in isolation from past farmland booms and busts, whose financially mediated dispossession laid the groundwork in fundamental ways for what was to come.

My history begins in the early twentieth century, but first it is important to note that we are discussing stolen land. All US land was originally territory of indigenous peoples—from the Haudenosaunee and Mvskoke of the east to the Cheyenne and Osage of the Great Plains to the Chinook and Chumash of the West Coast.[4] The violent, racialized dispossession of these original inhabitants and the enclosure of their lands under settler colonialism are the foundation upon which this entire history rests.[5]

## Precursors: Farmland Market Participation of Insurance Companies and Banks, 1910s–1980s

Life insurance companies and commercial banks have long been involved in farmland markets in the capacity of farm mortgage lenders. Throughout the twentieth century, life insurance companies provided between 10 percent and 25 percent of total farm real estate lending, the long-term nature of their liabilities providing a good match for the long-term income stream provided by farm mortgages. Commercial banks, meanwhile, tended toward shorter-term agricultural loans, but also contributed a sizable amount of financing for farmland purchases, generally ranging between 5 percent and 15 percent of the farm mortgage market between 1910 and 1990. Since the 1990s, insurance companies have gradually become less prominent players in farm mortgage lending, a response to financial stress and a broader shift toward more liquid assets;[6] banks, on the other hand, have greatly expanded their lending.[7] Government programs like the Farm Credit System

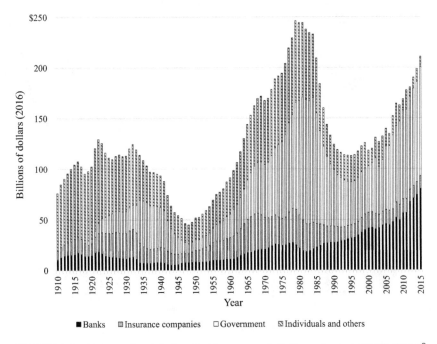

**FIGURE 1.1**  Farm real estate lending by source, inflation-adjusted, 1910–2015[8]
*Source:* Image by author based on data from Melichar, *Agricultural Finance Databook* (for years 1910–1959), and US Department of Agriculture Economic Research Service, "Farm Income and Wealth Statistics" (1960–2015).

and Farm Service Agency, as well as individuals, supplied most of the remaining farm mortgage lending. Figure 1.1 shows how the two main private institutional investors in farmland—banks and insurance companies—contributed to farm mortgage lending over the course of the century. Their combined farm mortgage lending still dwarfs their equity (ownership) investments in farmland.

The exceptions to this rule of financial engagement via mortgage lending occurred in tandem with the two great twentieth-century boom-bust cycles in US farmland markets (see figure 1.2).[9] During these periods, financial institutions were sometimes tempted into making equity investments in land, attracted by the rapidly rising land prices of the booms or the rock-bottom deals of the busts. The busts also left lenders the inadvertent owners of a great many foreclosed farms.

The first major farmland boom of the twentieth century began in the second half of the 1910s. US farmland prices increased rapidly, fueled by wartime demand for agricultural exports, low interest rates, and speculative fervor.[10] However, this boom was followed by a bust in the early 1920s as European agricultural production recovered from the war and the Federal Reserve raised interest rates in an attempt to curb inflation; the US entered the "roaring twenties" with farmers struggling to hold on to their land.[11] Though farm incomes recovered later in the decade, land values would remain low for a long time to come.

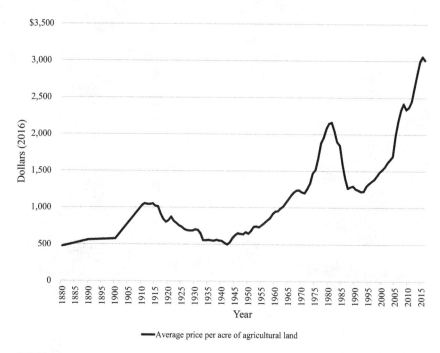

**FIGURE 1.2**   Average price of US agricultural land, inflation-adjusted, 1880–2016[12]
*Source:* Image by author based on data from US Department of Agriculture National Agricultural Statistics Service, "Quick Stats."

The 1920s farmland bust coincided with a period of widespread experimentation with large-scale farming techniques. Historian Deborah Fitzgerald recounts how corporate farming companies, driven by a belief in technology and scientific expertise, attempted to introduce methods of industrial efficiency into agriculture. This period saw the emergence of "chain farms," groups of properties controlled by a financial institution or professional farm management company. The chain farm owner or manager made the important decisions about what crops would be grown and what machinery would be used, though tenant farmers generally carried out the actual farm operation.[13] Some financial institutions took advantage of the low farmland prices of the 1920s to buy a portfolio of farm properties, which they operated at arm's length using the chain farm model. The Bluffton Bank of Indiana, for instance, established the Central Investment Company, which acquired nineteen farms comprising 2,900 acres between 1923 and 1927, each operated by a company employee. The goal of this company was to buy dilapidated farms; make improvements to soil, buildings, and farm management; and resell them at a profit.[14]

Financial institutions' accidental acquisition of foreclosed farms also contributed to the development of chain farming.[15] Caught up in the euphoric boom mentality, lenders had become lax in granting mortgages during the 1910s, and so many became landowners in the 1920s whether they wanted to or not.[16] The insurance company Aetna, for instance, owned six hundred midwestern farms by 1931. These were all rented to tenant farmers under the supervision of a farm manager.[17] Farm management companies emerged that specialized in managing farms on behalf of absentee landlords. One such company, Doane Agricultural Services, founded in 1923, offered a service it called "liquidation management," which involved making rapid improvements to the foreclosed farms owned by banks and life insurance companies to prepare them for resale.[18] By propelling financial lenders into landlordism, the farm bust fed a fledgling industry of professional farm management services.

The Dust Bowl and Great Depression of the 1930s put a damper on institutional investors' nascent interest in farmland ownership. Even private farm mortgage lending sagged for a time, as gun-shy farmers refrained from taking on debt.[19] However, in the 1940s, land values began to increase again and continued to do so—slowly but steadily—throughout the 1950s and 1960s. Insurance company and bank farm mortgage lending began to climb again in the 1950s as confidence in land markets was restored.

In the 1970s, this plodding increase in land prices once again transformed into a mad dash. The US entered another farmland boom, this time lasting from roughly 1972 to 1981. The causes of the boom were many: global droughts in 1972, a huge sale of grain by the US government to the Soviet Union in the same year, the devaluation of the US dollar, a highly inflationary environment that translated into low real interest rates, and a certain amount of ungrounded optimism.[20] Exhorted by their government to "plant fence row to fence row" and fearing that ever-increasing farmland prices would soon put landownership out of reach, US farmers took out mortgages to buy land at formerly unthinkable prices.[21] Average national farmland values quadrupled in the course of the decade from 1971 to 1981.[22]

As farmland prices climbed, financial institutions were once again tempted into buying farmland on their own accounts. Two *New York Times* articles published during this period capture farmland's growing appeal to institutional investors and, with the benefit of hindsight, clearly profile a bubble in formation. The first, published in April 1977, bears the title "Solid Gold Down on the Farm," a hint that farmland was now valued for its financial rather than its productive properties (see figure 1.3). In it, the author marvels that farmland prices continue to climb steeply and independently of farm incomes, citing cheap credit and lack of confidence in the dollar as possible explanations for this land-buying fever. The second, published in November 1980, just months before the bubble burst, bears the title "Farmland Lures Institutions." Although the article pointed out that

# Solid Gold Down on the Farm

**By JERRY FLINT**

DAKOTA CITY, Iowa—"It scared me when I heard this land was for sale at $3,000 an acre," said Marvin O. Bacon, taking a break from spreading nitrogen on his cornfield near here. "But if you're in the farming business you've got to own land. And if you're going to get anywhere you've got to take chances."

So a few days ago Mr. Bacon took a chance: He bought these 80 acres of Humboldt County land, paying $3,900 an acre—a total of $232,000, mostly borrowed. This gives him 400 acres of his own Iowa corn land on top of the 800 acres he rents from others.

About the same time Dale Madsen, another Iowa farmer, was taking a similar chance. Sitting in his camper at Storm Lake with Eunice, his wife; Marie, his mother, and Ed, his father, Mr. Madsen gulped hard and signed a contract to buy 480 acres at $2,300 an acre—more than $1.1 million. He is trading off 160 acres of his own land to help pay for it. This leaves him owning 700 acres and in debt for $800,000.

What amazes bankers, investors and economists is not the fact that farmers are buying land—it's the high prices they are paying. The farmland boom is five years old, and it has really picked up momentum since 1975. The heart of the boom is in the Middle West, where corn and soybeans thrive. In the year ended Feb. 1, farmland prices jumped 38 percent in Illinois, 32 percent in Indiana, 35 percent in Iowa and 31 percent in Ohio, the Agriculture Department reports. And these figures represent farmland that will remain farmland—not acreage to be transformed into shopping centers or housing lots.

Although some farmland is bringing more than $3,000 an acre, the average price is much lower. The 36 percent gain in Illinois, for example, brought its average price per acre to $1,450, well below the top prices being paid in the state. This is because the average includes land that is not under cultivation, such as pasture and woods.

The farmland boom is broadly based. Outside investors as well as farmers are rushing to buy prime acreage. These outsiders include not only Americans but also foreigners.

"They are the old rich families, disenchanted, nervous or just scared out of their minds, depending on where they come from—Germany, France or Argentina," says Reed Oppenheimer, who sells land to foreigners from his New York headquarters. The minimum investment he handles is $1 million.

The foreign investors are looking for security, says Mr. Oppenheimer, who graduated from Yale with a history major. "I have clients who have had land in their family for 1,100 years. They say, 'When Napoleon came, when the Prussians came, when the jewels were sold, we survived as a family because of the land.'"

The foreigners and other big outside investors are not exactly welcome in the farming region. The Continental Illinois Bank and Merrill Lynch, Pierce, Fenner & Smith recently announced a plan to set up a $50 million mutual fund to invest in farmland. A great howl of protest arose in the rural Mid-

vest, and the plan collapsed. "Those farmers didn't want to wind up bidding against Merrill Lynch," said one agent for outside investors.

Bob Bergland, the Secretary of Agriculture, notes that about 60 percent of the farmland purchased has been for enlarging present farms. Sole proprietorships and partnerships accounted for 80 percent of all farmland transfers in the year ended March 1, 1976. But economists observe that in a tight market—and the present market for prime farmland is tight—a

growing demand from outside investors can push prices still higher even if farmers themselves buy the bulk of the land.

One banker predicts that, despite the failure of the Continental-Merrill Lynch plan, outside money will continue to move into farmland, creating a wider separation between absentee landowners and the farmers they rent the land to. Such a separation will actually be necessary, he insists, because the price of land will become too high for most farmers to buy.

The present boom means more than just a new crop of big land owners in the Corn Belt. The boom also accelerates the trend toward fewer and larger farms, makes it harder for young persons to get started as farmers unless backed by parents or in-laws and makes farmers vulnerable to any agricultural depression.

The number of farms in the nation declined steadily from 4 million in 1960 to 2.8 million last year, according to

*Continued on page 5*

Oliver Williams

**FIGURE 1.3** Newspaper article about rapid farmland appreciation and growing investor interest in land, *New York Times*, April 24, 1977
*Source: New York Times* and Oliver Williams (illustrator).

there had so far been "more talk than action" and that farmland holdings by institutional investors were still comparatively tiny, it noted that insurance companies were now making equity investments in land for their own portfolios. Prudential Insurance Company, the article reported, had increased its farmland holdings from 110,725 acres to 587,615 acres over a period of just two years. Phoenix Mutual, meanwhile, had added an additional $5 million in farm properties that year alone, bringing its total to $15 million invested in twenty properties.[23] Both articles also reported on the formation of professional asset management companies that aimed to make equity investments in farmland on behalf of a rapidly growing class of institutional investors: pension funds. As the following section will describe, this development was initially met with public outrage.

Once again, the belief that farmland prices could sustain their phenomenal ascent turned out to be an illusion. By the 1980s, the agricultural and financial factors fueling the boom had turned in lockstep to destroy it. On the agricultural front, a major increase in global production led to sinking farm incomes. At the same time that US farmers were bringing tens of millions of new acres into production and increasing yields at the behest of their government, European wheat farmers and South American soy farmers were also greatly increasing production in response to high prices. US farm exports fell steadily from 1980 to 1986. Changes in the financial environment were equally damning. In 1979, the Federal Reserve opted to attack inflation via a drastic interest rate hike. The federal funds rate rose from 10 percent to 20 percent over the course of a year, successfully reducing inflation. While good for the overall economy, the new low-inflation, high-interest-rate environment was devastating for US farmers. Land lost its appeal as a store of value, contributing to falling land prices at the same time as farm mortgage payments became impossibly high. The high interest rates also caused the dollar to rise in value, further depressing US farm exports.[24] During the 1980s, farmer after farmer lost their land in a wave of foreclosures that devastated rural America. The number of US farms had already been steadily declining for decades—from a high of 6.8 million in the mid-1930s to well under 3 million by the mid-1980s, with black farmers suffering particularly high rates of attrition due to rampant discrimination on the part of government service providers, as well as other forms of institutional racism.[25] Now this trend was suddenly exacerbated by unpayable debts, leading some farmers to suicide and many to bankruptcy.[26]

As land prices fell, so did investor interest in owning land, and financial lenders once again found themselves becoming the inadvertent owners of far more land than they had planned. The insurance companies were particularly hard hit because they had begun making aggressive loans toward the end of the 1970s, when land was already overpriced, and because they had tended to make more loans to large, highly leveraged agricultural operating companies that went bankrupt more quickly than their small and medium-size counterparts.[27] Insurance company land acquisitions during this period were primarily inadvertent—the result of their lending activities, not planned equity investments. At the height of the boom in 1979, the farm real estate held by life insurance companies was still valued at a mere $241 million. The value of their holdings, however, had leaped to a whopping $1.6 billion by 1984 and $2.4 billion by 1989. This tenfold increase in value of landholdings occurred because of farm foreclosures and despite sinking land values throughout most of the 1980s.[28] A boom-bust cycle had once again shifted large amounts of farmland into the hands of financial institutions.

The acquisition of so much foreclosed farmland was not a welcome development from the perspective of farm lenders. During this period, interest rates had risen rapidly, making bonds twice or three times as profitable as the 4 to 6 percent cash return coming from farmland. Life insurance companies therefore worked to reduce the share of farmland in their portfolios. While the land was in their inventory, they generally leased it out to farmers rather than operating it themselves.[29] In the politically charged atmosphere of the 1980s, when small and medium-size farmers were desperately trying to hold on to their land, insurance company landownership was politically controversial; the companies were accused of mismanagement, of dumping land into already depressed land markets, and of selling to corporate interests at the expense of family farmers.[30]

As was the case in the 1920s, the wave of foreclosures that attended the farm crisis drove an expansion of professional farm management capacity as farm mortgage lenders suddenly became farm owners. By 1986, life insurance companies alone owned about two thousand farms, and three major insurance companies had acquired farm management firms to boost their internal farm management know-how. Prudential Life Insurance Company purchased Capital Agricultural Property Services, a farm management subsidiary of Northern Bank and Trust Company of Chicago; Mutual of New York purchased Duff Farm Management Service of Indiana; and Metropolitan Life purchased Farmers National Company of Nebraska, the biggest farm management company in the country.[31] John Hancock Mutual Life Insurance Company, meanwhile, funded the development of Pacific Coast Farms, which would later become Farmland Management Services and be integrated into Hancock's agricultural investment group.[32] The experience insurance companies gained with managing portfolios of farm properties during the 1980s would, as we will see, make them natural leaders when farmland investment took off again three decades later.

What can we take from this history? First, financial institutions have not always had much interest in owning farmland. Until the late twentieth century, the US financial sector's experience with farmland investment mostly took the form of farm loan provision. Commercial banks and life insurance companies generally invested in land *indirectly*, via mortgage lending to farmers, rather than *directly*, by buying farmland themselves. However, a second important lesson from this history is the enormous upheaval brought by land market booms and busts. Booms have always increased the attraction of farmland, spurring established farm lenders to dabble in farm ownership. Busts have had still more drastic effects, prompting a sizable transfer of land from farmers to lenders. Though financial institutions' interest in landownership quickly receded again in their aftermath, these boom-bust episodes left behind a subtly altered investment landscape; each gave rise to new types of farmland investment intermediary, paving the way for future rounds of investment.[33]

One project, conceived during the boom of the 1970s, threatened to break with the traditional relationship between the financial sector and agricultural land markets, foreshadowing changes that would come to fruition decades later. It is worth pausing to consider this project in some detail because of the historical continuities and discontinuities it reveals; it shows that the idea of investing institutional capital into farmland has been brewing for decades but that its political palatability has altered drastically.

## Ahead of Its Time: The Case of Ag-Land Trust, 1977

In 1977 the agricultural boom was ongoing and farmland values were increasing at an impressive, if slightly unnerving, pace. In the previous year, land prices had risen 17 percent nationally and a jaw-dropping 33 percent in the Corn Belt.[34] An article that year in *Forbes* magazine enthused about a new investment fund, Ag-Land Trust, which aimed to harness these impressive returns on behalf of pension funds: "A logical long-term hedge against just about anything that can possibly go wrong in the world ought to be farmland. Why not a mutual fund investing in such land? Why not indeed figured Merrill Lynch and Chicago's Continental Illinois National Bank, who are readying what they call AG-Land Fund I. If the fund sells out, you can be sure there will be a Fund II and III and so on, plus countless imitators."[35] The fund under discussion did not, in fact, sell out. It was canceled a month after the article's publication without having raised a cent.

Ag-Land Trust proposed to introduce the benefits of farmland investment to a new class of institutional investors. Shares would be offered only to pension funds and profit-sharing trusts, both of which are exempt from capital gains tax, meaning that the fund, too, would be tax-exempt. The fund hoped to raise a total of $50 million, with each investor owning shares worth between $100,000 and $5 million and committing to an investment term of at least five years. The capital raised would be used to purchase US farms, diversified by geography and crop, which would be leased out to tenant farmers. According to the declaration of trust, the purpose of the fund was "to invest in *United States agricultural land primarily for appreciation and secondarily for current cash return* to the Participants."[36]

The Ag-Land Trust proposal came at a time when corporate participation in agriculture was under political scrutiny. Since the 1930s, the number of farms in America had been declining steadily, while the average farm size grew. Fearing that the windfall agricultural profits and booming land values of the 1970s would be the nail in the coffin of the American family farm, several states enacted laws that restricted or prohibited corporate ownership of farmland, corporate participation in farming activities, or both. Oklahoma, Kansas, Minnesota, Wisconsin, South Dakota, Iowa, Missouri, and Nebraska all put into place some kind of "anticorporate" farming law during the 1970s.[37] These laws sought to uphold a vision of American agriculture centered on the land-owning family farmer. South Dakota's law discussed "the importance of the family farm to the economic and moral stability of the state," while Minnesota's law aimed "to encourage and protect the family farm as a basic economic unit, to insure it as the most socially desirable mode of agricultural production."[38] Though financial institutions were not the only target of these anticorporate farming laws, several explicitly restricted the ability of pensions or trusts to own farmland or required them to resell farms acquired through foreclosure within a predetermined period of time.[39]

In this atmosphere of heightened concern about the fate of the family farm, the Ag-Land Trust proposal provoked an immediate outcry. It met with contempt and fervent opposition from farm-state politicians as well as civil society organizations representing farmers, environmentalists, consumers, and religious organizations. Though insurance companies and banks had sometimes made equity investments in farmland in addition to their mortgage lending practices, this seemed to be a different animal: an investment vehicle explicitly designed to attract financial investors into farmland markets. The opposition crystallized in late February 1977 with three days of hearings before the Subcommittee on Family Farms, Rural Development, and Special Studies, of the House of Representatives Committee on Agriculture. Almost fifty people gave statements at this hearing, and all of them, except the representatives of Ag-Land Trust, gave negative evaluations of the proposal, some going so far as to suggest that federal legislation should be created to ensure that no proposal like this could ever get off the ground.

Criticism at the hearings centered on the probable effect such a fund would have on land prices. "The threat of a trust like Ag-Land," said Representative Paul Findley, a Republican from Illinois, "is that with its virtually unlimited capital resources, it will always be able to pay top dollar for land. Pitted against a mammoth Chicago bank, the small farmer will never win the bid to purchase quality farmland."[40] Several members of Congress expressed concern that financial investment in farmland could create a vicious cycle of investment and appreciation; if Ag-Land Trust was successful, other financial actors would soon follow its lead, driving up the price of land still further and enticing yet more funds to join the fray. Some went so far as to suggest that this was the intention behind the proposal. As Representative Harold Volkmer, a Democrat from Missouri, put it in a written statement submitted to the committee: "It appears that the financeers [*sic*] in Chicago and New York have seen agricultural land increase in value and have seen a way to step in and cause further increases after purchasing a large quantity of land themselves."[41]

Such criticisms reveal a shared (and surprisingly bipartisan) vision about the appropriate scope of financial activity. Many of the politicians present implied that the role of finance was to supply capital directly to producers, not to use that capital to buy up productive assets for themselves. In his questioning of the fund representatives, the chairman of the committee, Democratic representative Richard Nolan from Minnesota, professed to be greatly disturbed by this aspect of the proposal: "You are withdrawing investment funds from private industry at a time when there is an acute capital shortage in this area of our economy, according to some of our best economists. This has the effect of bidding up the price of scarce agricultural property."[42] Others suggested that pension funds should maintain the financial sector's traditional relationship with farmland markets and use their capital to grant farm mortgages. "I respectfully suggest," said Representative Keith Sebelius, Republican from Kansas, "that if our Nation's pension fund[s] would like to invest in American agriculture they should follow the example of the major insurance companies, and make direct real estate loans to farmers at current rates of interest."[43] By buying land directly rather than lending money to producers, financial institutions were seen to be overstepping the bounds of suitable activity.[44]

Much of the criticism leveled at Ag-Land Trust also emphasized the unique value of landownership to America's farmers. Many representatives spoke about the singular importance of landownership to their farming constituents, to the success of US agriculture, and even to the American way of life. In Sebelius's words: "When the farmer endures hard times, it is the knowledge that he is the steward of his own soil and master of his own fate that gives him the determination and fortitude to tough it out. Individual farm ownership has been the catalyst for the success of American agriculture throughout the world."[45] These comments evoke the "agrarian ideal," which has long accorded special political and moral status to farmland ownership in the US.[46]

In their effort to defend themselves, the promoters of Ag-Land Trust also emphasized their reverence for the family farm, though they presented the interests of the farmer in a different light. Ag-Land Trust, they argued, would in fact help family farmers by relieving them of the expensive burden of landownership and freeing up farmers' capital for operating expenses. Charles Hall, executive vice president at Continental Illinois National Bank, argued that to the small farmer "land ownership in and of itself is less important than the profit of his overall operation. In fact, the leverage provided to his operation by the ownership of highly capitalized land in other hands is of substantial economic advantage to him."[47] Ag-Land Trust, in short, would allow farmers to focus on the lucrative operation of the farm without having to trouble themselves with the capital outlay associated with landownership. This, according to the president of Merrill Lynch Hubbard Inc., James Mooney, was just another means of facilitating the flow of capital to producers: "We view our participation in the Ag-Land Trust as a logical extension of Merrill Lynch Hubbard's traditional activities of bringing additional sources of capital—in this case pension funds—to those who have need of capital and have difficulty obtaining it through conventional sources."[48] These comments reframe the fund's activities from monopolizing land for financial gain to supplying capital to those who use land as a productive asset, just as the insurance companies and banks had always done.

These arguments went over like a lead balloon. Representative Findley vehemently objected to the suggestion that farmers would be pleased to free up working capital by leasing rather than owning land:

> I cannot think of a single constituent of mine who has a farm who would say that ownership of the farm is less important than profit. I would say the same thing about those who are leasing farms. There is nothing that lies so deep within the soul of a farmer as the desire to own his own land. I think that is really the crux of the matter and is why such an outpouring of sentiment has occurred.
>
> My suggestion would be that you go back to the drawing board and drop the idea. You really have been tinkering with the virtue of country America with this proposal. I think that is very plain from the discussion we have had today. Ownership of land is of great importance to the farmer. He prizes it above almost anything else in his life.[49]

That Ag-Land Trust is here construed as "tinkering with the virtue of country America" makes explicit what was implied throughout the hearing: that landownership by financial institutions rather than farmers constituted a moral threat. Its critics did not reject it on the grounds that it was illegal or unprofitable, but rather that it broke with traditional ideas about how the financial sector *should* behave. The representatives of Ag-Land Trust had unwittingly launched themselves into an ethical struggle over the meaning of land, finance, and the American way of life. It is little wonder that, shortly after weathering this congressional storm, they withdrew their proposal.

Around this time, however, the role of the financial sector in the US economy was on the brink of changing in ways that would ultimately alter the possibilities for financial penetration of farmland markets.

# Finance Expands Its Reach

Over the course of the 1980s and 1990s, institutional investors grew to a previously unimaginable size, as individuals increasingly entrusted their savings to pension funds, mutual funds, and other professional money managers.[50] As these financial heavyweights searched for investment opportunities, they steadily expanded the purview of financial activity. Within the agri-food system, this occurred most noticeably in markets for agricultural commodity derivatives. These markets, which have existed in the US since the mid-nineteenth century, were traditionally dominated by farmers, food processors, and other "commercial traders" seeking to hedge against volatility in the prices of agricultural goods. Corn futures, for example, were a way that the farmers who sell corn and the mill owners who buy it could agree on a sale price in advance and plan their business activities accordingly. Financial actors—known in commodity markets as "noncommercial traders"—also used derivatives to speculate on commodity price movements for profit, but their participation was circumscribed by long-standing US regulations that allowed federal regulators to place "position limits" on noncommercial traders, restricting how many derivative contracts they could hold at a time. These regulations aimed to allow just enough speculation to enhance liquidity and assist with "price discovery," but not enough that it might start distorting food prices.[51] In other words, finance was legally subordinated to the needs of producers.

All this began to change, however, in the 1990s, when financial investors started speculating heavily in commodity markets through over-the-counter (OTC) investment vehicles—those traded outside of formal stock exchanges. Money flowed into OTC products such as commodity index funds, which are designed to track the price movements of a bundle of different commodities.[52] The investment banks that sold these investment products began making the case to regulators that they needed commodity futures to hedge the financial risks of selling these products and that they should therefore be considered commercial traders. In response, the Commodity Futures Trading Commission (CFTC) started, in the early 1990s, to grant exemptions to the position limits faced by these banks, essentially reclassifying them from "noncommercial traders" to "commercial traders" and thereby dismantling the conventional barrier that had existed between hedgers and speculators.[53] The newly unfettered commodity index funds swelled—from approximately $15 billion in invested capital in 2003 to $200 billion by mid-2008—as the bursting of the dot-com and other market bubbles sent institutional investors searching for new investment outlets.[54]

For scholars with roots in Marxist analysis, the financialization of the global economy is closely linked to the crisis tendencies of capitalism. The rise of finance was precipitated in the 1970s by a severe "crisis of overaccumulation" within American capitalism—the masters of industry had built up enormous amounts of capital, but now economic stagnation was making it hard for them to find profitable investment opportunities. As a solution, they turned away from investments in production and increasingly put their money instead into financial transactions, causing a "financial explosion," in the words of Paul Sweezy and Harry Magdoff, or a "phase of financial expansion," in Giovanni Arrighi's more temperate terminology.[55] However, this shift from productive to financial sources of profit did not resolve the problem of overaccumulation, but rather continuously displaced it, as an influx of investor capital into one asset class after another led to a series of bursting bubbles—the US savings and loan crisis of the 1980s, the Asian financial crisis of 1997, the bursting of the dot-com bubble in 2000, the subprime mortgage crisis of 2007, and many more. The never-ending search for investment outlets results in a proliferation of financial products, but also in capital pouring into the physical environment. In what economic geographer David Harvey terms a "spatial fix," excess capital flows into the built environment (such as infrastructure and real estate development) or into new geographic areas where expanded markets or new sources of raw material provide a temporary outlet.[56]

One way to absorb excess capital is through urban development—as evidenced by the neighborhoods of abandoned McMansions left behind by the subprime mortgage crisis; another is by buying large tracts of rural land. Beginning in the 1980s, therefore, at the same time as institutional investors were pouring money into financial assets such as agricultural commodity swaps, they were also acquiring large tracts of US timberland. The financialization of timberland has been thoroughly documented by Andrew Gunnoe. Historically, Gunnoe tells us, the US forest products sector—which makes wood, paper, and other tree products— was composed of vertically integrated enterprises; the same companies that produced the paper also owned the trees, and owning large amounts of timberland was considered essential to their success. But financialization changed the definition of success. By the 1980s, as the pressure to generate shareholder value grew, the enormous amount of capital that forest products companies had tied up in timberland holdings began to look like deadweight on their balance sheets. That money—the logic went—should be working for them, rather than just sitting in the ground. As a result, the forest products corporations began to look for ways to sell or otherwise monetize their timberland holdings. At the same time, the Employee Retirement Income Security Act (ERISA), passed by Congress in 1974, along with similar laws passed by several states, encouraged pension funds to diversify their investments beyond the bonds and other fixed-income securities that had historically dominated their portfolios. These investors now looked at timberland and saw a tantalizing alternative asset class.[57]

These factors combined to produce a gradual transfer of timberland owner-ship from forest products companies to institutional investors throughout the 1980s and 1990s. Like farmland, timberland was attractive to investors as a port-folio diversifier and inflation hedge. Unlike farmland, however, it was already consolidated into large corporate landholdings, allowing institutions to buy in bulk. In short, both farmland and timberland were ripe for financial takeover, but timberland was the lower-hanging fruit. During the 1990s, timberland owner-ship was also particularly profitable as the government tightened restrictions on logging in old-growth forests to protect endangered species.[58] Institutional inves-tors generally acquired timberland in one of two ways: either it was managed on their behalf by a timberland investment management organization (TIMO), or it was included in a real estate investment trust (REIT). Over three decades, TIMOs and REITs came to control a combined 36 million acres of US timberland.[59]

## Grandfathers of the Farmland Space: Farmland Investment Management in the 1980s and 1990s

As the financial sector's scope of action grew, even dowdy farmland became the target of tentative investor interest. In addition to a steady flow of investor capital into timberland, the 1980s and 1990s witnessed the creation of several companies and funds dedicated to facilitating equity investments in farmland by institu-tional investors. Though a farmland investment project proposed in 1980, just three years after the failure of Ag-Land Trust, faced a similar political backlash to its predecessor,[60] by the late 1980s the strident political opposition to financiers buying US farmland had mostly dissipated. This likely had two causes: first, that after a decade of explosive financial growth, people had grown accustomed to the pervasive presence of finance; and, second, that the economic devastation of the 1980s farm crisis further sapped the already declining political clout of farmers.[61]

Thanks to the farmland management capacity it had gained during the farm crisis, the insurance industry was at the forefront of this new farmland invest-ment activity. Insurance companies saw an opportunity to proffer their farm management services to pension funds, which were rapidly emerging as the new financial heavyweights. This impulse to offer financial services—to find more ways of making money by managing money—was very much in accordance with the ascendant spirit of financialization. By the early 2000s, there were four major US agricultural asset management companies—one asset manager I interviewed referred to them as "the grandfathers of the farmland space" (see table 1.1).[62] Three of these "grandfathers" had evolved out of the life insurance industry: Hancock

**TABLE 1.1** Four major US agricultural asset managers

| COMPANY | ASSETS / ACRES MANAGED | KNOWN INVESTORS | APPROACH | BRIEF HISTORY |
|---|---|---|---|---|
| Hancock Agricultural Investment Group (HAIG) | $3 bn 336,000 acres (as of 9/2018) | Alaska Retirement Management Board, Dallas Police and Fire Pension System, Florida State Board of Administration, New Mexico Educational Retirement Board, Orange County Employees Retirement System, Park Street Capital, San Diego Employees Retirement System, Western Conference of Teamsters | Farmland in US, Australia, Canada. More heavily invested in permanent crops than other asset managers (over 50% of properties managed). Some properties leased out while others are operated by vertically integrated or third-party management firms. | In 1981, John Hancock life insurance company offered its first fund, which included farmland and timberland debt and equity. In 1990, HAIG was founded and created a straight farmland equity investment product. HAIG is now a wholly owned subsidiary of Canadian insurance company Manulife Financial, which bought Hancock in 2004. Farm operations in the US are managed by integrated subsidiary Hancock Farmland Services and in Australia by Hancock Farm Company and Hancock Farmland Services Australia. |
| Prudential Global Investment Management (PGIM) Agricultural Investments | $986.7 mn unknown acreage (9/2018) | Nine institutional investors since 1989 | All US farmland. Includes row and permanent crop properties, some leased out and some self operated. Specific details are unknown. | PGIM is owned by Prudential Financial, a financial services company that began as a life insurance company. In 1986 Prudential acquired the farm management and brokerage company Capital Agricultural Property Services, and Prudential's agricultural investments team has been managing third-party institutional capital since 1989. |

| | | | |
|---|---|---|---|
| UBS AgriVest | $1.4 bn unknown acreage (10/2017) | Alaska Retirement Management Board, Army and Air Force Exchange Service, Iowa Public Employees Retirement System, Orange County Employees Retirement System, Sonoma County Employees Retirement System, Town of Manchester, CT, Pension Board | All US farmland. Target allocations of 60% commodity, 20% vegetable, and 20% permanent. Leases out properties. | AgriVest spun off from Connecticut Mutual life insurance company in 1983. In 1989 it was acquired by financial management company Batterymarch, became Batterymarch AgriVest, and began investing for institutional clients. Acquired by Swiss financial services company UBS in 1999 to become UBS AgriVest. |
| Westchester Group Investment Management | $8.2 bn (10/2018) over 2 mn acres (12/2018) | AP-Fonden 2, British Columbia Investment Management Corporation, Cummins UK Pension Plan, Environment Agency Pension Funds, Greater Manchester Pension Fund, New Mexico State Investment Council, New York State Common Retirement Fund, TIAA general account | Farmland primarily in Australia (40%), Brazil (38%), and the US (13%), but also Poland, Romania, Chile, and New Zealand. Roughly 95% of the holdings are in row crops (mostly grains, oilseeds, and sugarcane). Row crop properties are leased out, permanent crop properties are custom farmed. | Westchester was an independent farm management company founded in 1986. Its joint venture with Cozad Asset Management (Cozad/Westchester) began buying farmland for institutions in 1989 or 1990. In 2010, TIAA acquired majority ownership of Westchester Group. Now named Westchester Group Investment Management, the company manages properties globally on behalf of TIAA's global investment manager, Nuveen. The capital invested comes from both TIAA and other institutional investors. |

*Source*: Created by author based on data from various sources. HAIG: Hancock Agricultural Investment Group, "About Us," "Farmland Management," "Company Overview"; HighQuest Partners and Koeninger, "History of Institutional Farmland Investment"; Oakland Institute, *Down on the Farm*. PGIM: Prudential Global Investment Group. "Agricultural Equity": Schneider, "As More Family Farms Fail." UBS AgriVest: Jacobius, "More Investors Turn to Farmland"; Oakland Institute, *Down on the Farm*; HighQuest Partners and Koeninger, "History of Institutional Farmland Investment." Westchester: Nuveen, "Westchester Group Investment Management," TIAA-CREF, "TIAA-CREF Announces $3 Billion", Nuveen, *Responsible Investment 2018*; Dodson, "Cultivating Opportunities."

Agricultural Investment Group (HAIG) has roots in John Hancock and is now owned by the Canadian insurance company Manulife Financial; Prudential Global Investment Management's (PGIM) Agricultural Investments team was and is owned by Prudential Financial; and AgriVest (now UBS AgriVest) was founded by managers from Connecticut Mutual. We might think of these three as "FIMOs"—farmland investment management organizations—because of similarities to the TIMOs that bought up America's timberland during the 1980s and 1990s, and because all three of these parent insurance companies also sired TIMOs.[63]

The fourth "grandfather," Westchester Group Investment Management, has a slightly different backstory. Westchester was founded by Murray Wise, a farmland broker and property manager who authored books during the 1980s and 1990s about the benefits of farmland investment, making him something of a prophet.[64] In the 1980s Wise teamed up with Dale Cozad, an equity manager in Champaign, Illinois, who mostly invested the savings of University of Illinois professors. Cozard/Westchester began by creating limited partnerships to invest these individual savings into midwestern farmland, but eventually began attracting pension fund clients, starting with the Illinois State Board of Investment in 1989.[65] In 2010, Westchester Group was acquired by TIAA (then TIAA-CREF), giving it enormous capital resources and placing it at the forefront of the farmland investment industry's transformation. Today, Westchester falls under TIAA's global investment manager, Nuveen.

These asset managers offer institutional investors one or both of two options for investing in farmland. The first is the *separately managed account*, a customized portfolio of land assembled to suit the client's needs and owned, via a holding company, by the client. This service requires very high minimum investments—generally $50 million—meaning that it is accessible only to institutions and the very wealthiest of individuals. The second is the *commingled fund*, in which capital from a number of investors is pooled and used to collectively acquire farms. In exchange for either of these services, the asset manager charges fees, which may be negotiated on a case-by-case basis in the case of managed accounts. These four older agricultural asset management firms tend to have a relatively conservative investment strategy; UBS AgriVest is particularly low risk, buying only mature farmland in the US and leasing out its properties, while HAIG and TIAA take on slightly more risk by operating their own permanent crop properties and investing abroad (though primarily in relatively safe countries—Australia and Brazil). Historically, all have tended to take a long-term perspective on farmland, viewing agricultural properties as a way to generate regular rental income and increase the value of invested capital rather than something to flip for a quick profit.[66]

In the early years, convincing institutions to invest in farmland was not always an easy task. "There were times," one farmland investment manager told me of the late 1980s, "when we would call pension funds to try and arrange an appointment to even kind of tell them the farmland story, and they wouldn't even call you back." Another told a similar story about trying to sell the idea of farmland in the 1990s: "I'll tell you, the whole idea of institutional or nonlocals investing in farmland, I'd say the first fifteen years of us doing it people thought it was crazy. We knew more about farming than anyone on the island of Manhattan, but everyone else in Manhattan thought that that was something that no one should ever be proud of." Attracting institutional clients was never easy, and after a brief period of heightened institutional interest in farmland during the late 1980s and early 1990s, a booming stock market cast agriculture into even deeper shade. Phil, a former life insurance company executive who was running a US-based farmland management firm in the 1990s, described the thankless task of trying to recruit investors:

> And you'll remember, from '95 to 2005, what happened was that the equity markets were just booming, right? You could invest $100 million with a phone call and, given how tech stocks were doing and everything else, you could have a 16 percent return on them in six months. And so it was easy, highly liquid, volume was there. We'd go in marketing farmland, and this is an 11 percent industry, about half income, half appreciation, right? So we'd go in marketing an 11 percent return, very limited liquidity, and people would say, "Gee, it's beautiful, thank you. I like driving through it, but I don't want to own it when I can get all of this return somewhere else."

Farmland was simply not making high enough returns to compete with the stock market. As will be discussed in the following chapter, financial institutions, by their very nature, measure land according to financial criteria, assessing it for the returns it brings in. To them, its attractiveness is entirely relative, and during the "roaring nineties"[67] it still looked relatively unattractive.

In sum: from the late 1980s through the early 2000s the US farmland investment industry existed but had yet to go mainstream. A group of professional money managers—many of whom gained their experience managing foreclosed farms for the insurance industry during the farm crisis—were ready and waiting to invest pension fund and other institutional capital into farmland. The idea of institutional investment in productive rural land was gaining traction via timberland, and investment in farmland no longer faced such strong political opposition. But despite these favorable conditions, investor demand was simply not there.

# Searching for "a Big Huge Hunky Chunk of Land": Farmland Investment Goes Global

By the mid-oughts, however, things were changing once again. Beginning around 2005, agricultural commodity prices ramped up, bringing farmland prices with them. Then the 2008 financial crisis changed the entire calculus around farmland investment; faced with flagging stock prices, investors were suddenly willing to accept much lower rates of return. Many simply wanted a safe haven for their capital. Timberland, as a more established real asset class, would have been an obvious choice, but it was experiencing slumping profits as a result of the housing crisis, which reduced demand for sawtimber and devalued timberland on urban fringes formerly slated for development.[68] Farmland was a natural alternative. As farmland finally became a hot-ticket item again, the venerable asset management companies described above, as well as a handful of newer ones, were ready and waiting. Institutional investors, eager to buy farmland while it was still cheap, turned to these established companies, and the capital they controlled (their assets under management—AUM—in finance-speak) began to swell.

The number of US farms owned by institutional investors began to grow. The National Council of Real Estate Investment Fiduciaries (NCREIF) has been collecting data on US farms owned by institutional investors since the early 1990s for its Farmland Property Index. As figure 1.4 shows, the number of farm properties in the NCREIF index climbed steeply from 309 at the end of 2005 to 796 by the second quarter of 2018. This is not a perfect data source; the index does not include every institutionally owned farm in the US, and the changing number of farms in its index may represent fluctuations in membership as well as farm sales or purchases by existing members. However, it is the most comprehensive data source available and a relatively good indicator of growing institutional interest in US farmland.[69] By mid-2018, the value of institutionally owned US farms in the index topped $9 billion, over $5 billion of it in annual cropland and slightly under $4 billion in permanent cropland.[70]

Suddenly in the early 2000s, rather than struggling to attract institutional clients, agricultural asset managers faced the opposite problem: they couldn't spend money fast enough. Their primary focus on US farmland limited their ability to absorb the flow of institutional capital into agriculture. Institutional investors think in big figures—tens or hundreds of millions of dollars—and that buys a lot of farms. As one North American farmland investment manager lamented in an interview, "You don't find farms in fifty-, seventy-, hundred-million-dollar pieces like you do office buildings." As a result, he explained, his company faces some frustrating math: The average, "bite-sized" US farm in their portfolios, he said, costs $5 million. If a client wants to invest $100 million, they must therefore

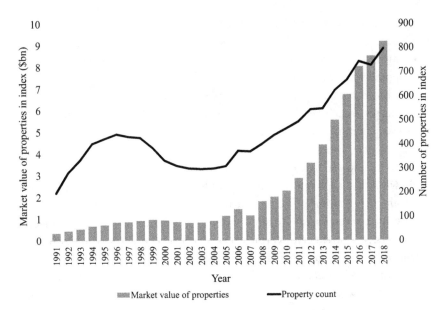

**FIGURE 1.4** Number and market value of institutionally owned properties in the NCREIF Farmland Property Index, 1991–2018 (Q2)

Source: Image by author based on data from National Council of Real Estate Investment Fiduciaries, "Farmland Property Index."

buy twenty farms. And for every farm the company buys, it inspects four or five others. The entire process is very labor intensive and time consuming.

The difficulty of shoehorning hundreds of millions of dollars of institutional capital into a tight US farmland market leads to frustrating delays for investors, which is particularly problematic given how quickly prices can rise during a boom. Take, for example, the experience of the Orange County Employees Retirement System (OCERS). OCERS embraced farmland investment with gusto, committing $80 million to a separately managed account with HAIG in 2011 and $40 million to a commingled fund managed by UBS AgriVest in mid-2012. When OCERS got on board, the UBS AgriVest Farmland Fund had almost $150 million in the "queue"—capital that had been committed by investors but not yet used by the fund. It therefore took over a year for the OCERS capital to be invested. The HAIG investment, meanwhile, was even slower off the mark. At an OCERS meeting in September 2015, HAIG managers reported that they had still invested less than half of the retirement system's capital. It was evidence, they argued, of their disciplined approach to investment that they were not buying overpriced farm properties that would subsequently perform poorly, even though they don't begin receiving fees until the capital is invested. While US-based managers are sometimes able to rid themselves of a big chunk of capital through an unusually large purchase—UBS AgriVest, for instance, paid $67.5 million for a 9,965-acre Wisconsin property in 2012—they are generally hemmed in by the time it takes to perform due diligence on scores of small farms, all while racing the clock of land appreciation.[71]

The mismatch between the scale of institutional capital (huge) and the scale of US farms (much less huge) is one of the main factors contributing to the global scope of the recent farmland rush. As the amount of institutional capital trying to enter farmland markets grew, some asset managers began to consider the need for a spatial fix; saddled with more investor capital than they knew what to do with, they contemplated finding fresh outlets for its deployment abroad. Phil—the life insurance executive turned farmland asset manager—explained that by 2005 his company had already begun to think about it: "I began to look around when we were getting, you know, a $100 million tranche of funds from a pension fund—it would take us a couple of years to invest it because it's a very limited market. You know, we can't go out and create what we want to buy. We have to take it when the market presents it to us. So we were struggling to invest many funds. And so I began to look globally and say, 'What would happen if we created a global investment candidate?'" The defining material characteristics of land markets—heterogeneity, localness, fragmented ownership, and relative illiquidity—all hinder the rapid deployment of institutional capital. And, of course, you "can't go out and create" what you want to buy—unlike many other commodities, land is not a manufactured good and has a finite supply. As a result, some existing farmland asset managers, like Phil, began to add foreign properties to their portfolios, while many more farmland investment companies were created with that explicit purpose in mind.

Going global facilitated institutional investment in farmland not only because it expanded the potential number of properties available for investment, but also because some countries offer possibilities for larger-scale purchases. Finding and acquiring land is still a time-intensive process, but it helps if the plot you end up buying is 20,000 acres, rather than a few hundred. Another US-based farmland fund manager explained that, in contrast to the US, where the government encouraged family farming through policies like the Homestead Act, creating a multitude of relatively small farm properties, Brazil's concentrated landownership structure offers the possibility of scale: "Brazil is entirely different than the United States. It is one where 1 percent of the population own 99 percent of the assets, and so that's very attractive to large institutions because they can come in and lay out 25 or 50 million dollars without too much work and have some big huge hunky chunk of land." Africa is another good place to find scale, since in many African countries the state is the sole landowner. In countries like Mozambique, the national government has the ability to grant thousands or even millions of acres to foreign investors at a time.[72] The search for institutional-scale properties helped push farmland investment from a relatively local or national business to a global one. With the exception of Westchester—which under the auspices of TIAA's asset management subsidiary, Nuveen, now manages land in the US, Australasia, South America, and Central and Eastern Europe—the "grandfathers" have confined themselves to highly developed farmland markets. But a new cohort of farmland asset managers has been more geographically adventurous.

## New Breeds of Farmland Asset Manager

As more financial capital began flowing toward the "farmland space," the handful of experienced farmland investment managers were joined by a throng of lively newcomers. Institutional investors control so much capital that their small collective shift toward farmland provided nourishment for many new farmland funds. Agricultural asset managers popped up like mushrooms on a giant log. Since wealthy individuals and family offices also became more interested in farmland, there were also new opportunities for asset managers to cater to smaller investors. These asset managers offer a broad spectrum of investment vehicles, including private equity funds, hedge funds, venture capital funds, and specialized agricultural funds operated by more mainstream asset managers. Between 2006 and 2016, over one hundred unlisted agriculture- and farmland-focused funds closed, raising $22 billion in combined capital (see figure 1.5). In particular, the

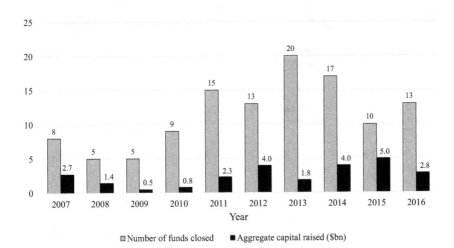

□ Number of funds closed     ■ Aggregate capital raised ($bn)

**FIGURE 1.5**  Annual unlisted agriculture/farmland-focused fund-raising, 2007–2016
*Source:* Author's adaptation of figure from Preqin, "Global Natural Resources Report."

years 2010 to 2014 ushered in a slew of new funds. Table 1.2 lists the biggest new agriculture/farmland funds as of 2016.

What these new funds offer is variation. Investors are diverse (see table 1.3), and so are their approaches to buying land.[73] There is a financial adage that asks, "Do you want to eat well or sleep well?" It captures the essential trade-off inherent to investing: some investments have high returns but also high levels of risk (they allow you to "eat well"), while others have low returns but also low levels of risk (with these you can "sleep well"). For most of the eighties, nineties, and early 2000s, farmland was primarily seen as a sedative, its gradual appreciation appealing to only the most risk-averse. Around 2006, as farmland markets began

**TABLE 1.2**   Largest fund managers by aggregate capital raised for unlisted agriculture/farmland funds, 2006–2016

| FIRM | HEADQUARTERS | TOTAL FUNDS RAISED IN LAST 10 YEARS ($BN) | NO. OF FUNDS RAISED IN LAST 10 YEARS |
|---|---|---|---|
| TIAA Asset Management | US | 5.0 | 2 |
| Paine & Partners | US | 2.1 | 2 |
| Proterra Investment Partners | US | 1.9 | 5 |
| Macquarie Infrastructure and Real Assets | UK | 1.2 | 2 |
| NCH Capital | US | 1.2 | 1 |
| Altima Partners | UK | 1.0 | 1 |
| Kenana Agriculture | Sudan | 1.0 | 1 |
| AMERRA | US | 0.6 | 2 |
| BRZ Investimentos | Brazil | 0.5 | 1 |
| Blue Road Capital | US | 0.4 | 1 |

*Source:* Preqin, "Special Report: Agriculture."

**TABLE 1.3**   Institutional investors in agriculture/farmland by type

| INVESTOR TYPE | PROPORTION OF INVESTORS (%) |
|---|---|
| Public pension fund | 20 |
| Endowment plan | 14 |
| Foundation | 12 |
| Private sector pension fund | 12 |
| Family office | 6 |
| Government agency | 6 |
| Asset manager | 5 |
| Investor company | 5 |
| Other | 20 |

*Source:* Preqin, "Special Report: Agriculture."

to heat up in many countries, land became attractive to both the eat-well and sleep-well crowds, but their different reasons for buying land necessitate different approaches to investment.

How investors view farmland—as a conservative investment whose purpose is to hedge inflation and diversify their portfolio, or as an aggressive investment whose purpose is to make them a considerable profit in a short amount of time— affects their approach to investment. Different types of end investors tend toward more conservative or aggressive farmland investment strategies. Pension funds tend to be relatively conservative because they are highly regulated and because their predictable, long-term obligations to retirees depend on a slow but steady stream of income. Hedge funds, on the other hand, are generally open only to accredited investors and are therefore subject to fewer regulations and can make riskier investments.[74] Yet this relationship is not set in stone, and an equally important factor in determining farmland investment strategy is the role that managers expect it to play in the investment portfolio.

There are five primary dimensions along which farmland investments differ (see table 1.4 for a stylized depiction). The first is ownership strategy. Investors looking to get exposure to farmland have two basic investment strategies at their disposal—I will refer to these ideal types as "own–lease out" and "own-operate." The *own–lease out* approach is generally thought of as more conservative but also less profitable. The investor simply acquires the land, finds a tenant farmer, and begins receiving an income stream in the form of rental payments, all while

**TABLE 1.4**  Attributes of farmland investment projects and level of risk and return

| RISK-RETURN | OWNERSHIP STRATEGY | INVESTMENT VEHICLE | COUNTRY | LAND TYPE | CROP TYPE |
|---|---|---|---|---|---|
| Highest | Own–operate | Direct purchase | Least developed (e.g., Africa) | Undeveloped land | Livestock and dairy Permanent crops |
| ↕ | | Private fund or operating co. | Intermediate (e.g., Brazil, Eastern Europe) | | Row crops (fruit and vegetables) |
| Lowest | Own–lease out | Public equity | Most developed (e.g., US, Australia) | Mature farmland | Row crops (grains) |

*Note:* This type of representation is fairly common in investment reports and presentations at investment conferences. For instance, some of these dimensions were also listed in a figure by Stookey and de Lapérouse ("Agricultural Land Investment," 4) and one by Hancock Agricultural Investment Group ("Company Overview," 4).
*Source:* Created by author.

gradually accruing capital gains from appreciation. The land acquisition and leasing is often done via an external asset manager, who in turn takes a cut of the profits. This approach appeals to investors who view land as a relatively long-term source of stable returns, portfolio diversification, and inflation hedging. In the second approach, *own-operate*, the investor is financially involved in both the purchase of the land and the agricultural production on it. In this case, the investor is exposed to the higher risks and returns associated with engagement in agricultural production itself, making it particularly popular among those drawn to agricultural investment for the potentially high profits.[75] Investors interested in agricultural production but not farmland ownership could also adopt a third approach, *lease-operate*, in which they produce on rented land, a strategy that certainly carries the highest risk but also has potential for high returns.

A second dimension along which farmland investments differ is the type of investment vehicle. Here the least risky approach, at least theoretically, is to buy farmland through the stock market because investors can choose how much to invest and the investment is completely liquid. However, the main way of doing this is to buy stock in a publicly listed agricultural operating company, which means that you are investing in agricultural production in addition to farmland. However, the advent of public farmland REITs—discussed in chapter 3—is now making it possible to buy stock in a company whose sole mission is farmland ownership. On the opposite end of the spectrum, the riskiest approach for an investor to take is probably to buy farmland directly, without the help of any intermediaries. However, for investors who know a lot about the investment location (such as wealthy individuals from rural backgrounds), this risk is much reduced, and the lack of intermediaries means no expensive management fees eating into their returns. In between these poles are a range of separately managed accounts, commingled farmland funds, and private operating companies whose levels of risk and return depend largely on their investment locations and strategies.

A third dimension of difference is the country within which investment takes place. Many of the factors that influence the desirability of an investment location are political and economic. Investors consider factors such as regime stability, government transparency, currency volatility, country credit rating, economic growth rate, trade policy, relevant taxes, availability of loans, agricultural subsidies, and farmland market liquidity. The weight of political factors with any given investors depends, again, on whether they hope to eat well or sleep well. One pension fund executive, for instance, told me that he would not consider investing anywhere in Africa out of concern that high profits might tempt the government to repatriate the land. However, this feeling is not universally shared. The executive of an agricultural company operating in Tanzania told me that he was contemplating a new project in South Sudan at the behest of one of his investors. He had a much more complacent perspective on political risk, saying, "South Sudan's eight months old, but it's as big as Texas. So imagine that Texas periodically had war with Louisiana. You wouldn't want to go anywhere near the Louisiana border, but if you were in the rest of Texas, it's as safe as anywhere else in Africa." A year after this interview took place, civil war broke out within South Sudan, which I imagine may have introduced an unacceptable level of risk even for him.

Though still disproportionately focused on North America, many new farmland/agriculture funds are either geographically diversified or have a regional focus in the Global South. Investors with an interest in the former Soviet Union and Eastern Europe, for instance, can invest with the asset manager NCH Capital. Its two funds, NCH Agribusiness Partners I and II, which closed in 2007 and 2014 respectively, together boast over 1.7 million acres of farmland in Ukraine, Russia, Bulgaria, Romania, and more.[76] Investors wishing to get exposure to African agriculture could instead buy into Emergent Asset Management's African Agri-Land Fund, which controls over 25,000 acres in Mozambique, Swaziland, Zambia, and South Africa.[77] Interested in Asian farmland? Duxton Asset Management, based in Singapore, has a globally diversified portfolio of 1.3 million acres of farmland, including operations in India, Sri Lanka, Vietnam, Laos, and the Philippines.[78] Farmland is now a truly global asset class. Figure 1.6 shows the geographic distribution of capital in agriculture/farmland investment funds—note, however, that this figure does not include investment via operating companies, which accounts, at least in part, for the seemingly low investment in Latin America.

A fourth factor in weighing the risk-return profile of a farmland investment is the level of development of the land. The lowest risk-return projects involve acquiring a fully functional farm property, while the highest risk-return projects require the wholesale conversion of forest or pastureland into farmland. In between is a full spectrum of improvements—often referred to as "transformations"—that can be made to a farm property before reselling. The more transformative projects are generally higher on the risk-return spectrum. The topic of farmland transformation will be picked up again in chapter 2.

Finally, different crops have different levels of associated risk. Permanent crops—perennials like grapes, tree nuts, and apples—have a higher risk than row crops, which include annuals like corn, soy, cotton, and vegetables. Whereas row-crop operations store value only in the land and infrastructure, permanent-crop operations also accumulate value in the vines or trees. Though they can be very lucrative, they are also more susceptible to major losses due to fire, weather, or disease. Additionally, while land appreciates over time, permanent crops depreciate and must be replaced. Livestock operations are also

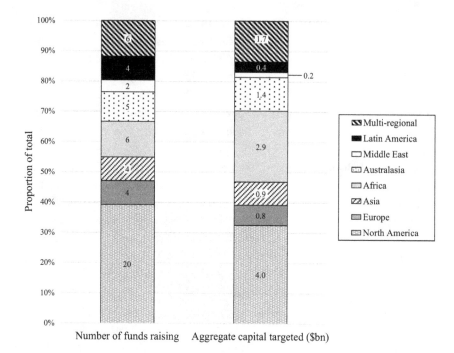

**FIGURE 1.6**   Unlisted agriculture/farmland funds in market by primary geographic focus (as of February 2017)

*Source:* Author's adaptation of figure from Preqin, "Global Natural Resources Report."

on the higher risk-return end of the spectrum for the same reasons. Within row crops, grains are lower risk than fruits or vegetables because they are less perishable, reducing the likelihood of spoilage and increasing the time that they can remain in circulation. In the US context, subsidized "commodity crops" are also lower risk because the government essentially guarantees a certain level of profitability.

Combining these factors leads to a wide array of investment options. So, for instance, an investor looking for a very low risk-return farmland investment might find a fund that specializes in long-term investments in US corn/soy land and that can arrange to lease out the land to a tenant farmer or operating company. On the opposite end of the spectrum, an investor hoping to make a 20 percent return will be willing to take on more risk. This investor can find a farmland fund or company dedicated to acquiring "undeveloped" land in Africa, turning it into a working farm, operating it for a few years, and then selling it. The scores of new agricultural investment funds created over the last decade combine these elements in different ways so that there is now a farmland investment product to appeal to almost any type of investor.

The financial sector's investments in farmland were mostly limited to mortgage lending for decades, followed by the plodding expansion of equity investment during the 1980s and 1990s and then a sudden blossoming in the early 2000s. The financial sector's growing interest in farmland ownership is partially attributable to the overall dynamics of financial capitalism—the concentration of capital in the coffers of institutional investors and their restless search for new asset classes in which to invest. But it has also been facilitated by the historical dynamics of farmland markets, including the earlier actions of financial institutions. In the US, the investment rush of the mid-2000s was, in important ways, a child of the 1980s farm crisis, whose debt-fueled foreclosures gave financial institutions experience in farm management that they would deploy when farmland prices resumed their rapid ascent two decades later. This history underscores that land booms and busts are engines of transformation—triggering both speculation and dispossession, they prompt the emergence of new market intermediaries and investment vehicles. However, the financialization of farmland involves more than just the appearance of new investment managers and funds; in the next chapter I argue that it also involves certain ways of thinking about the value of land.

# FARMLAND VALUES

I tell people that I come from a bond background [and] to me timber
is like a zero-coupon thirty-year bond. . . . Row crops are like a one-
year duration, income-bearing money market fund.

—Dave, US-based farmland investor

In 2010, the muckraking financial blog *Zero Hedge* reported on pension fund
TIAA-CREF's initial $2 billion allocation to agricultural land. In the comment
thread, readers attempt to parse the fund's motivations. One reader jokes that
a farmland bubble is emerging that will culminate with the appearance of a
new reality TV show: "*Farm Flippers*, Thursdays this fall on HGTV." The reader
even envisions some fake content: "Well the place was in total disrepair when we
bought it last month, but we cut down all the overgrown corn, painted those ugly
brown cows black and white, put lipstick on all the pigs, and of course put in all
stainless steel & granite feed troughs and watering buckets. We project we'll make
a 300% profit when we sell next month."[1]

This commentator assumes that institutional investors will bring a particular
financial mind-set to buying farms—one in which assets are valuable only for
the price they can fetch upon resale, not for their productive physical qualities.
Though couched in comic exaggeration (luxury décor and livestock cosmetics
are not, in fact, tools of the farmland-investment trade), this commentary nonetheless gestures toward an important aspect of the financial sector's newfound
interest in farmland: the way that financial landowners conceive of land's value.

The financialization of farmland has more than one dimension. It can be
seen in financial entities' growing participation in farmland markets, discussed
in chapter 1, but it also involves the diffusion of a certain way of valuing land.
The economic value of land has always involved a dance between its productive and financial qualities, but for finance-sector investors, the financial profits

that come from land appreciation take center stage. What's more, this manner of *thinking about* land is accompanied by particular ways of *talking about* land. Farmland fund managers, executives at agricultural operating companies, and other "farmland boosters" deliver investment pitches focused on the inevitable rise of farmland values with time, shaping investor expectations for the future and, with them, present land market conditions. Yet while investors value land largely for its financial qualities, the productive use of land remains essential to legitimizing ownership.

# Theories of Land Value

Economists have been struggling for centuries to pinpoint the source of land's value. Does the value of land come from its scarcity? From its ability to produce things that humans want? From some combination of the two? Revisiting classic and contemporary theories of land value can help us to understand why investors today are drawn to farmland.

## The Unearned Increment

The high point of land's economic career came in the seventeenth and eighteenth centuries with the preclassical French economic school known as the physiocrats.[2] Physiocrats, such as François Quesnay, believed that nature was the ultimate source of all wealth and that therefore only laborers who worked the land or the seas could create value. Farmers, miners, and fishermen were the "productive class" who generated value, while everyone else simply transformed or consumed the value that they had already brought to society.[3] This distinction between productive and unproductive activities—between value producers and value extractors—is what Mariana Mazzucato calls the "production boundary." This socially constructed dividing line, she points out, has regularly shifted with changes in economic thought and remains a politically significant issue today.[4]

The physiocrats were succeeded by the classical school of economics, which held sway from the late eighteenth through the mid-nineteenth centuries. Land no longer held pride of place at the center of classical economic theory, but it was still given distinct analytical treatment as one of the three "factors of production." The classical economists were extremely concerned with understanding two questions: first, how the factors of production—labor, land, and capital—interact to create value, and, second, how the income from production is distributed between those factors of production in the form of wages to workers, rent to landlords, and interest to capitalists, respectively. They placed labor at the center of value production and were fascinated and perplexed by land's seeming defiance of this "labor theory of value." How was it possible, they asked, that property owners could profit from land though they had done nothing to produce it and received its services at no cost? And—equally important—was it right that they should do so?

Most of the great political-economic thinkers at the time had their doubts about the legitimacy of land rent as an entire category of profit. Adam Smith, who is today revered as the original free-market thinker in economics, was notably critical of landlordism. In his 1776 tome, *The Wealth of Nations*, he spoke slightingly of landlords as the only class in society "whose revenue costs them neither labour nor care, but comes to them, as it were, of its own accord, and independent of any plan or project of their own." Unlike laborers (who earn wages through hard work) and capitalists (who earn profits by risking their capital), landlords make rent simply because they own land. This allows them to charge a "monopoly price" to anyone who uses it, a price generally far higher than what the landlord has laid out in improvements to the land. The lack of effort required for them to make money, Smith argued, too often makes landlords indolent and ignorant.[5] Though the production boundary had shifted since the physiocrats— now urban manufacturing was classed among productive activities—landlords were definitely still seen as value extractors rather than value creators.[6]

The ability of landlords to charge rent, the classical economists understood, was attributable to *both* land's usefulness in production and its scarcity. The Reverend Thomas Malthus put scarcity center stage, arguing that finite land resources, combined with a rapidly growing human population, would naturally lead to repeated cycles of poverty and starvation.[7] David Ricardo's influential theory of "differential rent," meanwhile, integrated land's productivity and its finiteness in explaining rent. Ricardo saw rent (and therefore land price) as ultimately deriving from what the land was able to produce, famously stating, "Corn is not high because a rent is paid, but a rent is paid because corn is high."[8] However, the precise level of rent paid to the landlord, he argued, was explained by land's scarcity and varying fertility; rent can be extremely high on the best-quality land in production and should hypothetically sink down to zero on the most marginal land.[9] For Ricardo, landlords were a feudal holdover, a parasitic class that siphoned money straight out of the pockets of the capitalist class and wasted it on luxury goods instead of recycling the profits back into production.[10]

For Karl Marx—simultaneously the greatest admirer and the greatest critic of the classical economists—the ability to collect rent was essentially about social power. He used the term "ground rent" to describe the portion of rent that comes purely from nature and not from buildings on or improvements to the land. Landlords' ability to collect ground rent, he argued, stemmed largely from the social power that allowed them to exclude others from their property and thereby extort a portion of the surplus value extracted by capitalists from workers.[11]

John Stuart Mill contributed to this discussion with the observation that rents—and therefore also land values, which are closely linked to rental income—tended naturally to increase over time, giving landlords yet another source of unmerited income. Since land is the only factor of production whose supply cannot be expanded, landowners automatically reap greater and greater returns over time as a simple result of economic and population growth. "The ordinary progress of a society," Mill argued, tends to give landlords "a greater amount and a greater proportion of the wealth of the community, independently of any trouble or outlay incurred by themselves. They grow richer, as it were in their sleep, without working, risking, or economizing."[12] Mill termed this automatic and ongoing appreciation in land rents/prices the "unearned increment" and argued that its accrual by landlords was fundamentally unjust.

Finally, no discussion of the political economy of land investment would be complete without a mention of Henry George. In his sensationally popular 1879 book *Progress and Poverty*, George, an American journalist, took Mill's view of land appreciation as the "unearned increment" and ran with it. Living in California in the late nineteenth century, George was surrounded by land speculation, due to everything from the Gold Rush to the construction of the railways. He was therefore acutely aware that the rent for the least productive land was nowhere near the theoretical zero that Ricardo had proposed, but was in fact greatly inflated by speculation. He describes how land speculators grow rich simply by acquiring land in a frontier boomtown. If you buy land during such a land boom, George states, "you may sit down and smoke your pipe; you may lie around like the lazzaroni of Naples or the leperos of Mexico; you may go up in a balloon, or down a hole in the ground; and without doing one stroke of work, without adding one iota to the wealth of the community, in ten years you will be rich! In the new city you may have a luxurious mansion; but among its public buildings will be an almshouse."[13] George blamed land rent for the persistence of poverty and rejected Malthusian doctrine as an attempt at shoring up landed-class privilege by pinning blame for hunger on natural scarcity rather than human-made causes. George politicized scarcity, describing famines in India and Ireland as the result of exorbitant rents charged by colonial landlords rather than the natural consequence of limited fertile land. Even more so than most of the classical political economists, George saw ever-increasing land rents and prices as a matter of social justice. He proposed a solution as well: a single, enormous property tax, which would replace all other taxes and be precisely calculated to absorb all of the rent earned by landlords, making speculation completely unprofitable.[14]

This classical political economic perspective insists that there is a meaningful distinction between the profits earned by agricultural producers on the one hand, and the rent and capital gains collected by landlords on the other. But during the 1970s and 1980s, as the scope of financial activity grew, a new generation of political economists argued that this distinction was becoming blurred. Surveying British landownership in the late 1970s, Doreen Massey and Alejandrina Catalano observed a growing number of financial landowners who, they argued, viewed their properties very differently than did England's traditional landowning class. Unlike the old, aristocratic landowners, pension funds and insurance companies viewed their landholdings as a source of future financial returns (in the form of rent and land appreciation) equivalent to any other investment.[15]

This trend toward thinking of land as a source of investment returns, rather than a foundation for productive activity, is one way to conceptualize the financialization of farmland. Like Massey and Catalano, David Harvey contends that investors are increasingly thinking about their land as a stream of future rental income indistinguishable in practice from the income stream they might get from their various financial investments. He argues that "for the buyer, the rent figures in his accounts as the interest on the money laid out on land purchase, and is in principle no different from similar investments in government debt, stocks and shares of enterprises, consumer debt and so on." He continues, "Under such conditions the land is treated as a pure financial asset which is bought and sold according to the rent it yields."[16] As land is increasingly treated as a "pure financial asset," Harvey argues, it no longer makes sense to distinguish between rent (the returns to landownership) and interest (the returns to capital). The boundaries blur as land becomes just one more way to extract financial returns from capital.

For Harvey, the treatment of land as a financial asset has some distinctly positive implications. It means that landowners are no longer simply barriers to capitalist progress—as Ricardo and Marx believed—but can instead play a useful coordinating role from the point of view of the capitalist system. In their relentless search for the highest rate of return on their capital, Harvey argues, financially motivated landowners force the capitalist producers who rent from them to put land to its "highest and best uses." Landowners who view their property as a pure financial asset will also be incentivized to make improvements to the land; rather than simply growing rich "in their sleep," as Mill believed, these landowners will perpetually chase higher rents—and thereby eventually higher land values—by making infrastructural improvements and generally developing land in accordance with its most profitable future use.[17] Harvey also acknowledges a darker side to the treatment of land as a financial asset, however, as it means subjecting land to all the irrational, speculative impulses associated with other branches of financial investment.[18] My research confirms that future increases in land rents and values (that unearned increment) are absolutely central to the investors' calculus. However, I highlight that the effort to convert land to its "highest and best uses" in pursuit of land price appreciation may itself be the cause of social and ecological harm.

## Land Value as Discounted Future Income

The neoclassical school of economics, which rose to prominence beginning in the twentieth century and still dominates today, provides a very different perspective on the value of land, and one that greatly influences investor decision making. Compared to their classical counterparts, neoclassical economists are less interested in explaining the production of value and more interested in understanding how markets function. Rather than competing social classes, their world is populated by rational individuals seeking to maximize their own well-being through market exchange. For neoclassical economists, value is no longer seen as an intrinsic quality of an object—an expression of its usefulness or of the amount of labor it embodies—but as a subjective quality determined by market participants. An object's value resides in its "marginal utility" to the buyer—its ability to satisfy the buyer's needs and desires at a particular point in time given how much of the thing the buyer already has. This marginal utility is revealed by the price that the buyer is willing to pay for the object in question on the free market. Only one type of value exists, and it is expressed by competitive market price. The words *price* and *value* have therefore come to be used synonymously. With this theoretical shift, neoclassical economists swept aside any moral questions about an object's "fair" or "just" value—questions that must naturally arise when value is considered intrinsic—making value a simple function of scarcity and demand.[19] They also moved the production boundary once again; now the supply of any good or service was considered productive activity as long as it had a price tag attached.

Today, under the influence of the neoclassical school, land is no longer seen as a unique category worthy of distinct analysis. Instead, it is treated like any other income-producing asset, whose value reflects the income it will bring in over time, discounted to present values. What follows is a *very* basic description of this income capitalization model.

In buying a piece of farmland, an investor will, first and foremost, consider how much he is likely to make in annual income from that property (either by charging rent or from farming—from the neoclassical perspective there is no meaningful difference between the two types of income). That level of income, divided by the price of the property, is the rate of return on the investment. In real estate investments, that expected rate of return is called the capitalization rate. The investor will buy the farmland only if the capitalization rate is comparable to or higher than the rate of return that he could receive from another investment with a similar level of risk. The capitalization rate essentially expresses the opportunity cost of putting money into this investment rather than some other money-earning investment or bank account.

$$Capitalization\ rate = \frac{Income}{Farmland\ value}$$

"Income" in this equation stands in for the *expected* returns to be made by farming the property or renting it out to a tenant farmer (we will return to the role of expectations in the next section). "Farmland value" refers to the price of land because, again, under current conventions value and price are seen as synonymous. This formula can be rearranged to give a basic model of farmland value.

$$Farmland\ value = \frac{Income}{Capitalization\ rate}$$

Here, the value of farmland is essentially the net present value of the expected future income stream from landownership, discounted to its present value at a risk-adjusted rate.[20]

This formula illuminates a couple of major reasons why land values might change. First, since farm income forms the numerator of the equation, farmland values will generally increase or decrease in tandem with farm incomes. So, for instance, rising agricultural commodity prices will increase farm rents and therefore land values. That much is relatively straightforward. Second, land values will also be affected by changes in the capitalization rate—the percent return that investors must expect to receive on their investment in order to bother buying land. Because it is in the denominator of the equation, increasing capitalization rates will cause land values to drop, and decreasing capitalization rates will cause land values to rise. Capitalization rates are very sensitive to changes in the prevailing national interest rate, which alter the baseline return that investors could make from putting their capital elsewhere. Low interest rates tend to boost farmland values (because low returns elsewhere make farmland more attractive to potential buyers), while high interest rates tend to depress farmland values (because investors who can get high rates of return just by parking their money in government bonds are less likely to make the riskier choice of buying farms).[21] This formula also explains why many investors view farmland as an inflation hedge; food is generally one of the first things to increase in price during a period of inflation, which may translate to higher agricultural incomes, while, at the same time, governments often combat inflation by keeping interest rates low, which means low capitalization rates.[22]

In reality, many other factors—such as tax policies and urban development opportunities—also go into determining the price of land. However, this model captures the most basic components, or *market fundamentals* that go into land-purchasing decisions: farm incomes and prevailing rates of return. This model puts flesh on the bones of Harvey's assertion that land is increasingly being seen as a source of future returns comparable to any other. Today's farmland investors, steeped in neoclassical economic thought, are always aware of how the returns on alternative investments are changing and how this may alter the profitability of buying land.

## Enter Irrationality

The income capitalization model is a useful rule of thumb, and it forms the rudimentary basis for how today's investors think about their own land-purchasing decisions. However, it also has some major shortcomings. It suggests that, since farm income is the main determinant of farmland price, the two should move roughly in tandem, making it very difficult to account for land booms and bubbles.[23] The neoclassical premise that individuals are guided by rational expectations and that market efficiency leads to accurately priced assets sets economists at a disadvantage where land is concerned. The dominant economic vocabulary also recognizes no meaningful distinction between speculation (betting capital on the belief that prices will go up or down in future) and investment (putting capital to work in a way that adds value and thereby ensures a future return), instead depicting speculation as just another form of investment.[24] Equipped with this conceptual tool kit, agricultural economists sometimes use terms like land-price "paradox" or "farmland-price puzzle" in reference to the confounding tendency for land to be priced above or below what it should be if everyone was operating with rational expectations about the fundamentals of farm income and interest rates.[25] Whereas farmland values often march along at roughly the levels predicted by dividing prevailing rents by the long-term interest rate, sometimes they shoot up to levels that rents and interest rates can't explain.[26]

Several material and social attributes make farmland markets particularly susceptible to booms and busts.[27] First, land is *finite*; more land cannot simply be produced to catch up with the level of demand, a fact that drives prices up and also provides a compelling rationale for would-be speculators. Second, decisions about land are subject to a *time lag*. A farmland purchase generally implies a commitment of at least one year—the length of one annual crop cycle—and much longer when it comes to building barns and other infrastructure. This difficulty in reversing course creates market inertia. Third, and most importantly, land is subject to *shared expectations*. The income capitalization model tells us that land values are based on expectations about market fundamentals; what it doesn't tell us is that there is no guarantee that these expectations will be rational. People form their expectations by reading the same government harvest projections and monitoring the same futures market indices. Shared beliefs also spread by word of mouth, with news of impressive profits spreading contagiously from one person to another.[28] There is, furthermore, a tendency to put too much weight on the experience of the recent past in establishing land-price expectations; after a few years of increasing land values, people begin to expect that prices will continue to increase indefinitely.[29]

This not-strictly-rational behavior of land markets is better explained by the theories of early twentieth-century British economist John Maynard Keynes and his theoretical descendants, such as Hyman Minsky. From their perspective, all markets are governed by expectations about a fundamentally uncertain future and are subject to the influence of unstable emotions. As a result, markets have been inflated by infectious optimism and speculative behavior since time immemorial—from the Dutch tulip mania of 1636, when a single tulip bulb could fetch as much as a house, to the dot-com bubble of the late 1990s, when investors poured money into overhyped internet businesses with names like Pets.com. Investors are emotional creatures, Keynes argued, who sometimes let their "animal spirits" get the better of them, leading to speculative episodes in which asset prices lose all grounding in actual asset values.[30]

While intoxicating optimism can infuse almost any market, land markets can also be fueled by economic anxiety. In times of economic uncertainty, people want to own something more solid than fiat currency. Paper money has no use value—if people stop believing in it, it becomes just so much ink-stained paper. Financial assets, likewise, can evaporate as fast as digital figures can change on a computer screen. But land is different. Its solidity and evident usefulness make it a safe haven that anxious investors can flock to when they fear for the value of their capital. Capital, as Marx observed, is generally governed by "laws of motion"; it can generate more capital only if it is in constant circulation, meaning that stagnant capital is considered a wasted opportunity.[31] However, this rule breaks down in times of economic uncertainty. At these moments, when investors are desperate simply to preserve the value of their capital, the fact that land is a scarce and appreciating resource becomes at least as important as its ability to generate profit by producing crops. At these times, investors are happy to take their capital out of motion and fix it safely in the ground. Land investment is therefore an inherently speculative endeavor, infused with optimism and anxiety, hope and fear.

## Cultivating Capital Gains

Do today's farmland investors view land as a "pure financial asset" akin to stocks and bonds? Do they treat it as a source of financial income rather than productive income, in line with the overall trend of financialization? Not entirely. In the aftermath of the global financial crisis, many investors were drawn to land precisely for its productive qualities. Many even chose to invest directly in farming instead of leasing out their newly purchased farms (i.e., they adopted the "own-operate" approach described in chapter 1). Yet, true to Harvey's description and in accord with the spirit of financialization, what unites farmland investors today—whether they treat land as a passive source of rental income or engage in farm operation—is that they all depend on the profits from land appreciation. It is this restless pursuit of capital gains, I will argue, that most characterizes the financial sector's turn to farmland.

Farmland investors conceptualize their returns as falling into two categories: the cash return and the capital gains. The *cash return*—also known as income return or simply the yield—to farmland investment comes from the regular income produced by the farm property; it can refer to either the returns from pro-duction or from rental income, if the land is leased to an operator.[32] The *capital gains*—also known as appreciation return—are the returns that come from increasing land value. For farmland investors who lease out their properties, returns are composed of cash returns in the form of rent, plus capital gains (all financial profits). For those who operate their properties, returns include cash returns in the form of farm income (productive profits), plus capital gains (financial profits).

For most investors, the cash returns from farmland ownership—whether in the form of rental income or farm income—would not alone be sufficient inducement to buy land. Cash returns on US farmland, for instance, are gener-ally in the mid-single digits, which is not enough to motivate most institutional investors. The relatively conservative pension funds frequently base their esti-mates of future obligations to retirees on a return expectation of between 7 and 8 percent,[33] while hedge funds often shoot for returns in the double digits. While investors are drawn to land's productivity and tangibility, they are not willing to make major sacrifices when it comes to returns. As a manager at one university endowment put it, "Farmland competes for every investment dollar [in our port-folio] like any other asset class would." In other words, investors are unlikely to temper their high return expectations to accommodate farmland.

Under these circumstances, modest farmland largely managed to capture the eye of capital markets during the recent land rush because of its potential to gen-erate capital gains. For pension funds, insurance companies, and other relatively risk-averse investors drawn to farmland as a long-term diversification strategy and inflation hedge, capital gains are central to the investment thesis; it is the reg-ular appreciation of farmland that allows it to store the value of invested wealth. For US farmland, many of the investors and managers I interviewed expected roughly 50 percent of their fund's total internal rate of return (IRR) to come from land appreciation.[34] One large midwestern farmland investor and operator described just such a 50/50 investment strategy: "The long-term historic returns to land appreciation have been between 4 and 6 percent in the core of the Midwest. . . . So that's why you invest, is that 5 percent appreciation per year. And then the second thing you get is, you should get, again, somewhere between 4 and 6 percent a year of rent." Despite his relatively staid approach to farmland investment, it is still the essentially speculative returns from capital gains that get top billing in his explanation. The capital gains ultimately explain "why you invest."

Farmland values do not always exhibit such impressive appreciation. Historically, the income returns for US farmland have stayed relatively flat, trending gradually downward, whereas the capital gains from appreciation have been far more erratic, going through booms and busts. Figures 2.1 and 2.2 show the source of farmland investment returns for two US states: Iowa and California. Figure 2.3, like figure 1.4 in the previous chapter, is based on data from the NCREIF Farmland Index of US farms owned by institutional investors, but rather than their total farmland assets (as in figure 1.4), it shows the breakdown of their returns. For this sample of institutionally owned US farms, land appreciation made up an impressive percentage of investment returns during the decade that lasted from 2004 to 2014, excepting a two-year lull at the height of the recession. It was only after a couple of years of these high appreciation returns—beginning around 2006—that institutional investors began considering farmland as a

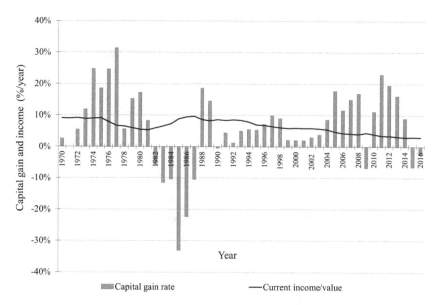

**FIGURE 2.1**   Return components for Iowa farmland, 1970–2016
*Source:* Bruce Sherrick, TIAA Center for Farmland Research, University of Illinois.

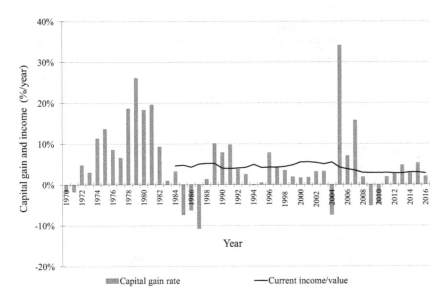

**FIGURE 2.2** Return components for California farmland, 1970–2016

*Source:* Bruce Sherrick, TIAA Center for Farmland Research, University of Illinois.

*Note:* Current income figures not available before 1984.

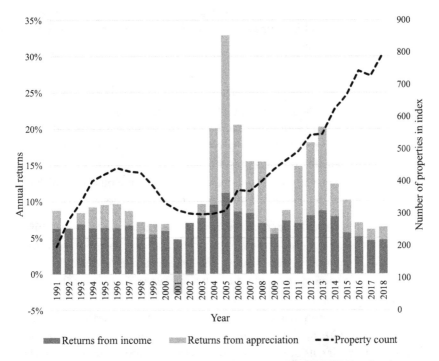

**FIGURE 2.3** Property count and return components for NCREIF Farmland Property Index, 1991–2018 (Q2)

*Source:* Image by author based on data from National Council of Real Estate Investment Fiduciaries, "Farmland Property Index."

serious possibility for their portfolios. Again, for many of these farms the income returns take the form of rent, meaning that they too constitute financial income; the appreciation returns are just the portion of the total returns that we *know* to be financial, rather than productive income.

Capital gains are, if anything, even more essential to the more aggressive farmland investors and asset managers buying land in "emerging markets." One approach is to buy land in regions that are undergoing particularly fast appreciation, whether due to policy changes, infrastructural improvements, or simply a booming land market. A participant in a conference panel on farmland investment in Latin America explained, "If you want exposure to commodity prices and a hedge against inflation, buy farmland in the United States, it's easier. If you buy into this concept of increasing pressure on farmland, buy some farmland in Latin America—you're going to get a large return and it's not going to come from the farming, it's going to come from the capitalization of the land." As with stock-market speculation, this kind of investment depends greatly on the timing of the investment.

Not all capital gains, however, come from passive appreciation. Many farmland managers stimulate land price increases through transformations that add value to their properties. Speaking at an investment conference, the CEO of a Brazilian agribusiness—one of several large operating companies now selling themselves as farmland investment vehicles—differentiated these two components of the capital gains:

> There are three types of returns, you know, in farmland. One is the operating return—1, 2, 3 percent, nothing to write home about, nothing to particularly attract private equity capital. Then there's the land appreciation. As you know, land does not depreciate over time, it appreciates over time at a rate which could go anywhere between 4 and 8 percent depending on the regression you run, the data you use. And then there's the land transformation return, ... This is a business which has generated 20 percent IRR over the past four, five years.

According to his math, the operating return barely signifies, while passive appreciation is considerable—and returns to transformation are potentially phenomenal. He is simultaneously speculating and investing in land-price increases. Though his company farms many thousands of hectares, the most important commodity it produces is the land itself.

There are many ways to add value to land. These include clarifying legal title and altering farm size to increase marketability. An important way to boost land's value is by increasing its productivity through major transformations—clearing forested land, making soil amendments, installing drainage tiles, drilling wells, or adding irrigation infrastructure. But agricultural yields, and therefore land values, can also be increased simply by employing more intensive production techniques, a fact that further blurs the dividing line between returns from operation and returns from appreciation. When I asked one European pension-fund manager why he preferred an own-operate approach to farmland investment, his answer was simple: "If you participate in the operating part of the business, you have better control over the land appreciation." For him and many others, productive use of land is, paradoxically, inspired by visions of capital appreciation.

Though increasing farm income—and the increasing land values that come with it—are most often achieved through conventional, input-intensive production techniques, there are exceptions to this rule. Farmland LP, a US-based farmland-management company that owns properties in Northern California and Oregon, increases land value by having tenant operators take land through the three-year organic certification process. Their strategy takes advantage of the fact that the premiums associated with organic certification are quickly capitalized into the value of the land, increasing its resale price.[35] Another example comes from Peoples Company, a farmland broker, manager, and investment service provider based in Iowa, which in 2018 began offering winter cover cropping to the investor-landowners whose properties it manages. They believe that this soil improvement initiative will appeal particularly to institutional investors because of their overriding focus on land price appreciation.[36]

The drive to increase land values has major socioeconomic and environmental implications. What almost all these approaches to "adding value" share is that they propel agriculture toward more capital-intensive, industrial forms of production. Higher yields are generally achieved through increasing use of technology and inputs. Value-added strategies relating to farm size, meanwhile, often involve consolidation of small landholdings.[37] Formalized land titles, too, often come at the expense of poor rural communities who, in many parts of the Global South, farm land to which they have no official use right. The land appreciation produced through these approaches also creates path-dependent outcomes; not only do high farm incomes lead to higher land values, but higher land values require farmers to produce more income in order to pay their rent—generally through intensifying production practices or producing higher-value crops. In her seminal work on the California organic industry, Julie Guthman revealed that even the conversion of conventional farmland to certified organic farmland may contribute to a "land value spiral," in which higher rents require more intensive production practices, a trend that favors wealthier farmers while allowing less time for soil and other biophysical systems to replenish themselves.[38] In other words, the "highest and best uses" of land from the perspective of capital accumulation are not necessarily the highest and best uses from the perspective of environmental sustainability or economic justice.[39]

Financial landowners value farms for their ability to produce capital gains as much as for their ability to produce corn or almonds or sugarcane. Many actively pursue capital gains through taking part in production or by converting land to higher-value uses. However, as the following section will argue, anticipation of "passive" appreciation is also a crucial part of land's appeal.

## Great Expectations

When farmland is valued as a stream of future income, the contours of that future become very important. As the income capitalization model shows, expectations of future farm incomes and future returns on alternative investments are key to determining farmland values in the present. This future orientation is not unique to farmland—within capitalism, market actors are always motivated by their own expectations of the future and also spend inordinate amounts of time trying to influence the expectations of others.[40] Stories are one of the primary vehicles used to shape market expectations, serving both to raise investors' "animal spirits" and to calm their anxiety.[41] Stories can also be incredibly effective at raising investor capital. Anna Tsing argues that businesses dependent on attracting up-front investment capital must whip up investor anticipation through fantastical stories of possible future profit, a process she describes as "spectacular accumulation." "In speculative enterprises," she argues, "profit must be imagined before it can be extracted; the possibility of economic performance must be conjured like a spirit to draw an audience of potential investors."[42] In other words, capital-hungry start-ups paint the picture of the profits that they hope will eventually materialize (and that, indeed, they will use investor capital to create). "Conjuring" future profit through spectacular narrative and images attracts investors, giving the company the capital and time needed to see if it can make that imaginary future profit into a reality.

Like other uncertain ventures, farmland investment depends on economic spectacle to convince prospective investors.[43] The stories told by farmland investment boosters can be seen as *performative*—their depictions of economic reality have the potential to shape investor behavior and therefore the very market outcomes they purport to describe.[44] By describing a coming farmland boom, they foment the investor excitement that could make that boom a reality. Attending investment events and conducting interviews between 2010 and 2014, when the farmland boom was in full swing, I found three future-oriented narratives to be particularly instrumental to conjuring investor capital and, with it, farmland price appreciation. The first two narratives directly influence investor expectations about the fundamentals of the sector, corresponding roughly with two key inputs of the income capitalization model of farmland value: farm income (rents or crop prices) and capitalization rate (the required rate of return). The third narrative, which concerns investment timing, addresses the potential problems that can arise from collective expectations. Here again, land's productive and financial qualities intertwine, with both instrumental to shaping visions of the future.

## The Scarcity Scare

Though the financial sector values farmland largely for its ever-increasing price (or exchange value, in rough Marxian terms), its productivity (or use value) nevertheless features prominently in the stories prospective investors are told about land. Scarcity narratives, first enshrined in economic thought by Malthus, are central to capital-raising efforts and therefore also to the construction of a farmland boom.[45] The particular incarnation of the scarcity narrative that haunts agricultural investment conferences begins with the fact that the Earth contains a finite quantity of land. This is followed by an enumeration of the factors currently leading human needs to outstrip food production: inexorable population growth, increasing meat consumption by the growing middle class in China and other emerging economies, and the advent of biofuels all contribute to an increasing demand for grain. Meanwhile, urbanization and climate change are daily reducing the stocks of available farmland. The aging farm population in the US, Japan, and elsewhere is occasionally reflected upon with foreboding. It is concluded that food and farm prices will increase into the indefinite future.

Though, during my research, the inevitability of resource scarcity was most often expressed in general terms, occasionally Malthus himself spontaneously entered the conversation. An American working on creating a Brazilian farmland fund told me that Malthus was making "a huge comeback," while a British fund manager based in Uruguay described increasing global resource scarcity as a "Malthusian perfect storm." At one major agricultural investment conference, attendees were given a DVD of Last Supper for Malthus, a peculiar French documentary on the global food crisis. The opening sequence jumps spasmodically between two-second clips—dark-skinned people dancing in traditional clothing, flies swarming on the face of a dead goat, a factory farm, a polar bear diving from an iceberg, a crowd of Asian people in suits, locusts swarming up a tree, a hurricane seen from space—all accompanied by sinister-sounding strings. The viewer is then introduced to the primary conceit of the film: televised interviews with the ghosts of Malthus and Ricardo, temporarily arisen to rehash their theories in light of the recent food crisis. Malthus argues that scarcity and population growth make starvation inevitable, while Ricardo contends that free trade and comparative advantage can serve to address scarcity. The film introduces a wide range of opinions through interviews with world food experts—including some who lay the blame for the food crisis at the door of biofuel subsidies and commodity speculation—but the overall effect is to reinforce the "scarcity scare."[46]

Some asset managers build their investment strategies around explicit visions of global resource scarcity. Such is the case with Jeremy Grantham, cofounder and chief investment strategist of Grantham, Mayo, and van Otterloo (GMO), a Boston-based asset management company that manages tens of billions of dollars. Grantham is an unapologetic neo-Malthusian and an enthusiastic proponent of farmland investment.[47] A quarterly letter he sent to investors in 2011 was a nineteen-page diatribe titled "Time to Wake Up: Days of Abundant Resources and Falling Prices Are Over Forever." In this letter, Grantham reiterates Malthus's argument that an increasing food supply leads to a population boom followed by starvation, and argues that the incredible wealth unlocked by the "hydrocarbon revolution" has led to just such a boom in the human population. He concludes that the result of this is both a dire scarcity of food and other natural resources and an excellent investment opportunity.

> If I am right in this assumption, then when our finite resources are on their downward slope, the hydrocarbon-fed population will be left far above its sustainable level; that is, far beyond the Earth's carrying capacity. How we deal with this unsustainable surge in demand and not just "peak oil," but "peak everything," is going to be the greatest challenge facing our species. But whether we rise to the occasion or not, there will be some great fortunes made along the way in finite resources and resource efficiency, and it would be sensible to participate.[48]

The letter demonstrates how the scarcity narrative is used to encourage investment in natural resources—given the inexorability and immediacy of the trends described, one would be foolish not to. GMO is not the only firm to base its investment strategy explicitly on resource scarcity. The Hamburg-based investment firm Aquila Capital, which manages funds for agriculture and other real assets, even has Dennis Meadows, the former director of the Club of Rome think tank and coauthor of its famous 1970s report about global scarcity, *The Limits to Growth*, on its board of directors.[49]

The scarcity narrative conjures a farmland boom by acting on one of the key elements of farmland value: expected farm income. Given that present land value is understood to be the net present value of future farm income, anything that causes expected farm incomes to increase will also boost present land prices. The story of rapidly growing food demand on a collision course with stagnant or shrinking supply is therefore performative; proponents of farmland and agricultural investment use it to conjure the investor capital and the investment returns that they need to be successful.

This narrative has consequences beyond the bottom line of farmland funds. As Henry George observed about Malthusian theory, an exclusive focus on natural scarcity diverts attention from the social injustices that actually lie at the root of hunger.[50] This critique has been elaborated by scholars like Amartya Sen and Michael Watts, who trace the ultimate cause of famines to economic vulnerability, not overall food shortages.[51] As Lappé and Collins put it, "There is scarcity, but it is not a scarcity of food. The scarcity is of people who have either access to the means to grow their food or the money to buy it."[52] The scarcity narrative obscures the social origins of food insecurity, and in doing so it legitimizes certain types of responses to events like the 2008 food crisis while cutting off others. If the food crisis resulted from insufficient global food supplies to feed a rapidly growing population, then recruiting investment capital to fund large-scale industrial agriculture in developing countries is a completely logical response. If, however, the problem is enormous economic inequality that results in some people eating concentrated grain in the form of Big Macs while others starve, or government policies that encourage burning grain as biofuel in gas tanks, or financial speculation that turns grain into a form of capital, then the solutions might be quite different. By obscuring the power relations that determine access to food, scarcity narratives depoliticize hunger and foreclose options for addressing it.

It is important to note that, for every agricultural investor who wholeheartedly gave me the Malthusian line, there was another who relayed it with substantial caveats. For instance, Leon, a hedge-fund manager I met at several agricultural-investment conferences, pointed out that this was not the first time scarcity had been a concern: "Populations were increasing in the eighties. And in the nineties [people said that] if everybody in China just ate one more, one egg a day, it would take all the grain surplus in the world." Markets, he concluded, sometimes fail to translate such rising demand into rising food or land prices. Interestingly, he doesn't take issue with the basic premise about population growth and resource limitation, but he does question whether this would inevitably lead to higher food and land prices. In general, the idea that demand for food is outstripping farmland supply is orthodoxy within the farmland investment community. The scarcity narratives that result have a performative effect on land values if they successfully mold investor expectations about future farm incomes.

## Financial Dystopia

Another common narrative in farmland investment pitches depicts looming economic crisis in Northern economies. In this narrative, financial uncertainty combined with mismanagement on the part of government—policies of quantitative easing and deficit spending are most frequently cited—will lead to a massive loss of wealth for those foolish enough not to have placed their money into real assets like land. While the scarcity narrative's applicability hinges on land's productive capacity, the financial disaster narrative relates primarily to its ability to store wealth.

The financial disaster narrative was impressively displayed by one of the featured speakers at the 2012 Land Investment Expo in Iowa.[53] The speaker argued that expansionary monetary policy and excessive government spending were paving the way for rampant inflation. "I haven't met one person in America who believes that Congress will do the right thing and begin to balance the budget," he announced. "It's a 100 percent–held consensus they will not do it. And if that's the case, it means they will continue to print a trillion and a half dollars a year, and when a government does that for a sustained period of time the value of your savings declines. Simple as that." He went on to explain that people are naturally turning to real assets as an inflation hedge, but cautioned that gold prices are too volatile to serve as an effective store of value. "I tell people, 'I understand if you don't believe government will ever do the right thing, I understand, but rather than do gold, why not buy farmland?'"

This speech not only captured the essence of the financial disaster narrative, but it also illustrated the centrality of spectacle and performance to the farmland boom. At one point in his presentation, the speaker casually reached back into his pocket and pulled out his wallet, saying, "In fact, a coin dealer gave me this bill the other day." He extracted a single bill and brandished it before the audience. "This is what people are afraid of," he told them. "This is why people are buying gold for sixteen hundred bucks. These are why people are buying cropland . . . because of this. This is what they don't want to see. This bill right here is the highest-denomination currency on the planet ever printed. It's from a country called Zimbabwe. And what happened there, they turned on the printing presses and never shut 'em off. This bill right here, this one piece of paper is a $100 trillion bill." He gestured up and down with the bill on the last few words, giving them extra emphasis. Laughter erupted from the audience. The speaker continued, "And it buys lunch today." Putting the bill back in his wallet, he explained its portent: "See, that's what your government can do. If they don't do the right thing. If they don't raise taxes and cut some spending, then all they'll do is print money and run bigger and bigger deficits and destroy your savings. End of story. So that's why I can say . . . cropland is perceived to be in a bubble, and at the same time say it's possible it could go up dramatically in value even further from right now."

This spectacular story of government ineptitude and hyperinflation feeds the fertile nexus between land value and fear. Land prices tend to increase when investors are afraid for the value of their money, and the image of the $100 trillion bill stokes these apprehensions adroitly. It connotes a future in which the relationship between money and time has unraveled. One of money's most essential attributes is its ability to store value, so that it can be earned in the present and expended on as-yet-unknown future needs and desires.[54] But in case money betrays our trust, farmland is proposed as a stalwart option for safely chaperoning capital from present to future. This narrative of monetary fragility conjures farmland appreciation by altering expectations about future capitalization rates. In the event of runaway inflation, interest rates will be extremely low, and (except for high farm incomes) nothing else could more surely contribute to high land values.

In addition to inflation, farmland is held out as protection against several other varieties of economic calamity. Lucas, the manager of a hedge fund with farms in Australia and Brazil, held pretty bleak visions of the economic future. In our 2011 interview, he explained,

> Most banks in the world are broke. Bankrupt. They are totally bankrupt. . . . You and most people on the street don't know that, but a lot of people sitting at the top know that, and they know that they need a tangible investment. So what is a tangible investment? It's a hard asset. Something that nobody can take away from you, in principle. . . . It cannot really, it can be, ultimately everything can be confiscated, but generally land is not confiscated. They might confiscate the currency, they might confiscate gold, but you might still have your farm.

He achieved the transition from discussing insufficiently capitalized banks to an extreme scenario involving government confiscations of private property in under a minute. Like the Land Investment Expo speaker's ominous vision of hyperinflation, this account marries economic and political insecurity in rendering a future in which private capital will not be safe. Here, the structural fragility of financialized economies becomes, somewhat ironically, a rationale for finance capital to flow into a new sector of the economy.

What is striking about the financial-disaster narrative is how vivid and alarming the potential future scenarios are. The keynote speaker at another investment conference made the argument for buying farmland on the basis that it is one of the few assets that could survive a currency collapse. "Let's say the dollar collapses," he told the audience. "I'm not saying it will, but let's say it does. If the dollar collapses, one of the few ways you'll be able to preserve your wealth and even make money is to own productive farmland." The intensity of these scenarios—hyperinflation, confiscation of private property, currency collapse—makes them more effective at conjuring farmland investment. The rhetoric of pending financial collapse—like the scarcity narrative—has a drama to it that seems almost certain to boost expectations. This narrative performatively shapes land values by influencing investor beliefs about the safety of their capital and the rates of return that will be available from alternative investments in future.

## A Tsunami of Money

Investment decisions made in the present are based on expectations not only about the uncertain future but also about the uncertain behavior of others in relation to that uncertain future, a phenomenon that Elena Esposito terms "circular uncertainty."[55] With temporal contingency compounded by social contingency in this manner, prospective investors are acutely concerned that they may be buying into farmland too late. Farmland boosters address this concern head-on through a narrative about investment timing. While the scarcity and financial disaster narratives work to conjure a farmland boom, this third narrative addresses concerns created by the boom itself. It reassures prospective investors that they are early rather than latecomers, and that they will be riding the surge of investment up rather than down. This narrative has been made largely obsolete by the cooling of land markets since 2014, but it is still worth examining for what it shows about the social nature of land markets.

The timing narrative begins by characterizing agriculture as a historically underinvested sector, but one that is teetering on the verge of discovery. According to a speaker at one agricultural investment conference, "There's a wall of cash that's about to hit agriculture." At another investment conference, a panelist used a different but equally evocative image: "There is a tsunami of money that wants to enter the agricultural sector." A farmland investment report, meanwhile, explains, "Current market conditions have created what might be described as a perfect storm in terms of buying opportunities."[56] These metaphors depict an impending boom that is extremely close, extremely large, and almost violent in its intensity. The implication is that *now* is precisely the moment to invest in agriculture. There are huge opportunities to be had as agriculture is brought into the twenty-first century—in the case of farmland, through a rapid increase in land price—but because this flood of institutional money is about to hit the sector, those opportunities won't last long.

That capital is pouring into land markets makes for a powerful selling point, but it also raises concerns about a possible bubble. To address these concerns, farmland investment promoters resort again to timing. Bubble or no bubble, they argue, if you get in soon, you can still make a profit before it bursts; the meteoric rise in farmland prices may no longer be in its earliest stages, but it still has a long way to go. A 2008 publication by the financial advising firm Wellington

West used a baseball analogy to describe this position: "We believe we are in the third or fourth inning of a significant farmland price appreciation cycle resulting from high farmer income combined with scarce arable land."[57] Three years later, a US-based farmland fund manager resorted to the same analogy in addressing the boom-bubble issue: "It's not the first inning of the game, [but] it's not the eighth inning either."[58] In a 2012 interview, meanwhile, a US-based hedge fund manager told me that we were in the fifth inning of the game. What is perhaps most noteworthy about this ubiquitous baseball metaphor is the tacit admission that the existing rate of farmland appreciation is a temporally bounded event rather than a sustainable increase based on secular trends (as the scarcity narrative suggests). The metaphor implies that the game *will* end at some point, but emphasizes that there is still plenty of time to play before then.

At a 2012 agricultural investment conference, an influential speaker argued that farmland prices had a long way to go up before coming down. After pointing out that agriculture was receiving unprecedented investor interest, he said,

> This too will someday end in a bubble. I know it's incomprehensible for some of you to believe that right now, but it may be in ten years, twelve years from now, everybody will be talking about agriculture. CNBC will become an agriculture-commodity TV network, and everybody will be talking about their farms and how much money they are making in soybeans and cotton. So, when those days come, when that bubble comes, I hope you're all smart enough to sell out. . . . But in the meantime, I urge you all to learn about farming.

Here a looming bubble is less a cause for caution than an inevitability, but the important message is that investors still have ample time to invest, make money, and sell out before the bubble bursts. The end will not come for a decade or more, and by the time it does agricultural investment will be pervasive. Surely, investors need have no fear about being too late to the game, if this is the future that awaits them.

This timing narrative, like the others, has its detractors. Quite a few interview participants (though none of those currently promoting a farmland investment project) told me that they believed farmland to be "topped out" or expressed concern about there being too much "hype" surrounding land. They were very aware of the social uncertainty created by collective investor beliefs and were determined not to fall victim to it.

The three investment narratives I have described may all influence future events, but that influence is not decisive. To say that these narratives are performative does not mean that their descriptions necessarily came to pass; it simply recognizes, as Judith Butler puts it, "that reiteration is the means through which

[an] effect is established anew, time and again."[59] Following the philosopher J. L. Austin, Butler argues that most performative discourse will succeed in influencing the future only under "felicitous" circumstances. Under infelicitous circumstances they may fail or "misfire," as Austin puts it, and the reality they describe will not take place.[60]

As it turned out, circumstances were infelicitous to the continuance of a farmland boom, though not definitively so. Falling agricultural commodity prices during 2014 and 2015 reduced the credibility of the Malthusian story line, while gradual economic recovery in many countries quelled economic anxiety. This change in circumstances wasn't enough to create a real bust, but it certainly took the air out of the farmland boom. There was no mad rush away from farmland or agriculture, but there was a quiet hissing sound. The (partial) misfire of these farmland investment narratives can be attributed in part to the same old agricultural risks that have plagued farmers for centuries—good weather contributed to rising supply and falling crop prices. But it also signals the financial risks involved in farmland investment—a booming stock market and the likelihood of rising interest rates also helped cast farmland back in the shade, while the inflation predicted to follow the expansionary monetary policies of the recession completely failed to materialize.[61] In addition, there are abundant location-specific causes for performative misfire; Oane Visser, for instance, describes how narratives about Russian farmland scarcity fell flat when it turned out that the supply of land was actually more than enough to satisfy investor demand without resulting in the hoped-for price increases.[62] Farmland investors are fundamentally future oriented, valuing land as a stream of future income, but there are limits to their ability to predict or influence that future.

## The Moral Economy of the Farmland Investor

Though land's financial qualities—its ability to produce a stream of future income in the form of rent and capital gains—are major enticements to investors, its productive qualities remain essential to legitimizing landownership. This is because land is encumbered by social values beyond the economic. Land is what Marion Fourcade terms a "peculiar good" and Margaret Radin calls an "incomplete commodity"; it is bought and sold almost everywhere, but because it is a means of production essential to human survival and livelihoods, its commodification is always partial and contingent, hemmed in by legal restrictions and social norms that limit the conditions of its monetization and sale.[63] According to Radin, incomplete commodification often occurs in objects that are closely linked to personhood; this includes the obvious—think human organs, children, and sex—as well as less controversial but still personal commodities such as employment and housing.[64] Farmland, too, has an intimate connection to self. It grows crops that, when eaten, are incorporated into our flesh, and it harbors personal cultural identity.[65] One could even say that land is "sacred," which is how Viviana Zelizer describes objects that society reveres and is therefore reluctant to debase with the cool calculation of market rationality.[66]

Finance, meanwhile, gets almost the opposite moral billing. As the branch of the economy that specializes in monetary transactions, finance must contend with cultural associations between money and the profane (think "filthy lucre" and "the root of all evil").[67] The US financial sector has worked hard to distance itself from associations with immorality, often by deploying alternative discourses about its work. Marieke de Goede recounts how, in the late nineteenth century, there was widespread distrust of the financial sector; many viewed stock-market investing as a parasitic enterprise, a means to gamble on the wealth generated by others without making a productive contribution to the economy. Over the course of the twentieth century, however, the financial industry waged a cultural, legal, and political battle to differentiate itself from gambling and craft a more virtuous image. It achieved considerable success. Financial professionals managed to improve their image by recasting themselves as experts in scientific risk management whose careful research and modeling are in themselves a productive contribution to the economy.[68] In Mazzucato's terms, finance worked hard to cross the production boundary, shifting in the public imagination—as well as in the calculation of national economic statistics like gross domestic product (GDP)—from value extractors to value creators.[69] Yet this wholesome image remains precarious, and financial professionals must perpetually shore up their shaky stance as virtuous economic actors.

In other words, *economic values* are not the only ones that matter; the *social values* associated with farmland present a challenge for the farmland investment sector.[70] Critics of financial landownership capitalize on the moral reputation of both farming and finance, depicting farmland as sacrosanct while casting those who wish to profit from it as profoundly unprincipled. Recall that, in the 1970s, the Ag-Land Trust farmland fund was characterized as "tinkering with the virtue of country America." It was accused of monopolizing precious farmland instead of leaving it to its rightful owners—the farmers—and of straying too far from finance's designated role of supporting producers by supplying them with capital. The fund promoters, meanwhile, defended themselves by arguing that they would aid agricultural production by freeing farmers from the considerable costs of landownership. After decades of financialization, this moral struggle is more muted but continues nonetheless, with the farmland investment community taking an active part in promoting a certain moral vision of their activities. Following E. P. Thompson and James Scott, we can think of their vision as a "moral economy"—a set of shared ideas about the rightful roles of different market participants and the fair distribution of scarce resources.[71] Despite their primary focus on the financial returns from landownership, the moral economy of the farmland investor is centered—somewhat paradoxically—on productive use of land.

## Becoming Farmers Overnight

At one agricultural investment conference in the early years of the boom, the moderator gave an opening speech in which he marveled at how much the topic of agricultural investment had increased in popularity. He joked, "We're having a little fun making history. We're pretty sure there have never been this many farmers at the Ritz-Carlton at one time," and received an appreciative rumble of laughter from the audience.[72] The choice to characterize the conference attendees as "farmers" may simply have been a matter of comedic license, but it also represents a common theme. When their biographies offered them any sort of choice in the matter, the asset managers and company executives I spoke with generally chose to identify themselves as farmers rather than as investors or entrepreneurs. This serves two purposes: positioning the speaker as an agricultural industry insider and assuming the positive moral connotations of farming.

In 2011 I interviewed Carlos, a senior executive at a Brazilian operating company that owns and operates thousands of hectares of farmland and sees farmland purchase and resale as a major part of its business. As we sipped espressos in a glass-walled São Paulo conference room, Carlos told me the story of how he broke into the world of finance. After earning a degree in agronomy and working as a farm manager, Carlos recounted, he decided to go to business school in the US: "To start with, it was pretty tough. . . . They would not accept me because I was a farmer. . . . I was the Brazilian farmer. Everybody knew me. I was this redneck in New York." He went on to explain that it took getting his MBA and becoming a partner at a major financial firm to be taken seriously despite his agricultural background. But now, he concluded laughingly, he finds himself in the opposite situation. When he recently went to New York for an investment conference, he found that "they're all 'Oh, you're a farmer, how interesting.' Now it's cool to be a farmer, but ten years ago, oh, it was a curse."

Carlos's story is about Wall Street's radically changing view of agriculture over the course of a decade, but it also performs a kind of "boundary work."[73] It delineates a strong boundary between two fields of expertise—agriculture and finance—and adamantly positions the teller on one side of that boundary.

With the story, Carlos situates himself as a farmer first and foremost, while his many years of professional experience in business and finance take a backseat. One function of this boundary work is to increase his credibility as an insider in a booming economic industry. Located at the intersection of agriculture and finance, the emerging farmland investment sector presumably requires expertise in both fields. However, in financial circles, business and economic knowledge are assumed, while agricultural knowledge is the new ingredient that needs to be demonstrated in order to attract investors.

Later in the interview, Carlos once again evoked the farmer-financier boundary, this time to disparage the competition: "You have some financial guys claiming to be ag experts 'cause the guy had an uncle who was a farmer in Iowa in the fifties. I used to live on a farm. I can claim that. I can ride a horse, I can castrate a bull, you know? I can do all that nasty stuff. I can drive a tractor. I did that." His emphatic assertion serves to discredit the competition by characterizing them as poseurs whose professional experience comes from the "wrong" side of the farming-finance boundary while simultaneously burnishing his own farming credentials (not to mention his masculinity). It is also an observation on the commonness of these efforts to appropriate the farmer identity. Since few agricultural investors can claim to have castrated a bull, they must emphasize whatever farming bona fides they can muster, even if it's just a farmer uncle.

This claiming of the farmer identity was a common theme throughout my interviews, as was belittling the competition as ersatz farmers. An executive at a German asset-management company with extensive farmland holdings also invoked the image of the poseur farmer: "[In 2008] some people, they were running around having discovered agriculture as a growth industry of the next ten years, who sort of went to Sunday school and became farmers overnight and produced all these wonderful glossy brochures and Excel spreadsheets and came up with fantastic return projections for plain-vanilla farming which were just not robust, not serious." The description of these money managers who "became farmers overnight" calls into question the competence of the competition. It characterizes them as outsiders—trend followers rather than trendsetters—and in doing so locates them on the wrong side of the boom curve. A US-based farmland fund manager made a similarly sarcastic observation: "There's lots of people that don't even know what they are talking about that have rushed into this area, that haven't farmed, that don't have any experience, that are now farming experts." Once again, this boundary work emphasizes the speaker's own presumably more authentic expertise while denigrating that of his competitors.

But laying claim to a farmer identity also serves another function: it lends to the speaker some of the positive moral traits associated with farming. By situating themselves as farmers, investors dodge the worst moral charge levied at finance—that of appropriating value rather than creating it—and associate themselves instead with the evident productivity of farming. This moral work came to the fore in a 2012 interview with Dave, a US-based farmland investor and investment manager with properties in multiple states. Dave expressed some serious concerns about the "fast money" that hedge funds were introducing into agriculture and suggested that pension funds were little better, calling them "wolves in sheep's clothing." Perhaps owing to his moral qualms about institutional investment in land, Dave repeatedly referred to himself as a "real farmer" throughout our interview. In the first minutes, he explained: "I'm a real farmer. I'm big enough that I don't run every tractor and plant every seed, but I go out there and put on my boots and work on the equipment and do stuff with the guys." Later, he reasserted this farmer identity when mentioning his past career in the financial sector: "While I'm a real farmer, I was also an investment banker and a Wall Street guy for [several] years—that's how I made enough money to go back to buying farms, because I didn't inherit any land, I just bought my own." His experience within the financial sector is constructed as a useful but secondary pursuit to his genuine identity as a farmer, although he in fact has considerable experience in both worlds.

The determination to be identified with farming rather than finance likely stems at least in part from an ongoing awareness of land's "sacred" properties. Though prizing land for its financial returns, the farmland investment community nonetheless pays homage to land's unique role as a means of production. By identifying themselves as farmers—if at all possible—they position themselves on the producer side of the production boundary, increasing the perceived legitimacy of their activities.[74]

## Transforming the Land

Even those farmland investors with no hope of being viewed as farmers can still position themselves as essentially productive actors by emphasizing their improvements to the land. The idea that productive use of land legitimizes ownership is deeply entrenched in Western thought, dating back to the English Enlightenment philosopher John Locke. In his "labor theory of property," Locke posited that when an individual mingled his labor with nature, he gained the right to subtract that piece of nature from the common resources of humanity, making it his own private property—an enclosure that was justified only as long as the land was put to use.[75] However, importantly, the owner needn't farm every acre himself or eat every apple he grows; Locke accepted money as a suitable, nonperishable way to accumulate the fruits of nature. In fact, just the act of making improvements to the land was, for Locke, a productive act. By transforming "wasteland," he thought, owners add to the overall wealth of society and legitimize their claim to the property. This argument was used historically to justify imperial territorial expansion; colonists argued that conquered lands were underused *terra nullius* (nobody's land), a claim that simultaneously erased indigenous people and justified the annexation of their territories.[76]

In accordance with this Lockean imperative to improve, agricultural investors frequently highlight the ways in which they add to and transform their properties. For example, Carlos was in the midst of describing his company's strategy of buying and reselling farmland when he paused to say: "Of course, we've got to realize that transformation means production. So we are not speculating in the sense that we're just buying land and sitting on it and waiting for something to happen." He wanted it understood that his company was actively adding value to land by investing capital in it. Emphasizing this point serves a practical economic purpose: it increases the company's appeal by stressing that it generates land price appreciation rather than waiting for unpredictable land markets to deliver. However, the statement "We are not speculating" also has strong moral overtones—it proclaims the righteousness of the investment. Unproductive use of land is a particularly sensitive and political topic in Brazil, as we shall see in chapter 4, but I found this emphasis on land transformation to be fairly ubiquitous (though generally not as explicit).

Emphasizing land transformation allows investors to position themselves as productive actors even if they are not actually involved in production and have no hope of being viewed as farmers. For instance, the manager of a US farmland private equity company that was still raising capital when I interviewed him in 2012 told me that his fund planned to create capital gains through land improvements and altered production techniques: "Instead of just being a *passive* owner collecting rent, we will be an *active* landlord who will rent the land out but will take an *active* role in increasing productivity through best practices, enforcement of those best practices with our tenant farmers, [and] land improvements that will improve the productivity of the land" (emphasis added). His company, he said, would transform its properties by changing farm boundaries, fixing drainage problems, and implementing no-till farming, and would increase yields by insisting that its tenants adopt the latest precision agricultural technology. These land transformations and technological improvements make the company an agent of transformation rather than "just" a "passive landlord." In this way, even completely rentier landownership can be located on the producer side of the production boundary.

The equivalence between farmland improvement and productive use is, of course, bolstered by the scarcity narrative. Against an assumed backdrop of global food scarcity, even investors who take no part in agricultural production can position themselves as productive (and therefore moral) actors by arguing that they are adding to the stock of farmland available to feed the world. Citing pressures from urbanization and desertification, an agricultural investment conference organizer described to me the urgency of bringing more farmland into production: "We've done the numbers, and if you look at it from an objective standpoint, and you model out supply and demand, you need to put something close to 70, 75 million new hectares of land into production. . . . There's this tremendous need." Some express the need to increase agricultural production in even more compelling terms. Lucas, the hedge fund manager who worried about banks failing, did not mince words on this subject either: "I am passionate about this. If we don't do something, we are going to die. It's that simple. And it's not going to be pleasant. It's going to be pleasant for me. I have my own farm. I can produce everything I need, but most people are going to suffer, and I hate to see people suffering. I mean, why? When we have so much. There's no reason for any of this to happen. Zero." The image of enormous and imminent human suffering lends an emotional power to the act of buying and improving agricultural land. It also casts agricultural investors in the role of savior. If they do their job well, they will benefit shareholders *and* stave off global hunger.

## Funding Farmers

Farmland investors also align themselves with agricultural production by arguing that their landownership frees up working capital for farmers to use.[77] In doing so, they position investor land purchases as an extension of the traditional (and therefore morally accepted) role of finance in supplying capital to producers. Farmers, in this framing, are more interested in expanding their operations than in owning land.

This perspective can be seen in full force in the online promotional materials put out by Bonnefield, a major Canadian farmland-investment company, which acquires its land through a "sale–leaseback" model, buying land from farmers and then leasing it back to the former owners on a long-term basis. The Bonnefield website provides case studies of farmers who sold to Bonnefield and prospered as a result. One begins: "For Jim and Lois Smith (not their real names), farming is in their blood." The case study goes on to describe the dilemma Jim and Lois faced as they wished to expand their operation and reduce their debt: "Jim's father was what you would call 'old school.' He was proud of his land and held onto it no matter the costs. Having grown up with that mindset, Jim and Lois were conflicted. To stay in farming, they would have to grow their land base but for their peace of mind and their children's futures, they also had to reduce their debt." Ultimately the story ends happily, with Jim and Lois selling 500 of their 800 acres to Bonnefield and using the proceeds to lease an additional 2,500 nearby acres from the company. By abandoning the "old-school" mentality of Jim's father and embracing a more modern perspective that privileged farm operations over sentimental attachment to land, the Smiths were able to reduce their debt, buy new machinery, and triple the area of their farming operation.[78]

The argument that farmland ownership is no longer a priority for farmers is based on a view of farming as an industry like any other. In an online interview, Bonnefield president and CEO Tom Eisenhauer describes the transfer of real estate and productive assets to institutional investors as an economy-wide trend: "This trend first started in commercial real estate back in the seventies and eighties. It's happened in hotels, commercial real estate, industrial real estate, even aircraft operators don't own their own planes anymore—they lease them. . . . Farming has been slow to adopt this. It's probably twenty years late, but it's a trend that we see beginning to happen."[79] That farmers are expected to follow in the footsteps of hotels and airlines flows logically from a view of farmers as primarily motivated by profit maximization and farmland as an asset class like any other; farmers, the Eisenhauer interview and other Bonnefield promotional materials repeatedly suggest, may wish to improve their "return on equity" by selling the farm and leasing it back. The assumption is that farmers, like financial landowners, view their land as just so much immobilized capital waiting to be set in motion. This positions investors to swoop in helpfully and take unwanted—or at least unneeded—assets off farmers' books, allowing farmers to do what they do best: farm. This perspective is not entirely inaccurate; an Australian study by Sarah Sippel and colleagues found that the financial logic of farmland investors resonated with some farmers who were "global players."[80] Indeed, David Harvey claims that "under conditions of capitalist landownership," landowning producers of all stripes will need to treat land as a financial asset in order to stay competitive.[81] At present, however, many farmers still have a pride in ownership that an enhanced revenue stream cannot entirely replace.

The moral economy of the farmland investor comes down, in essence, to this: investor ownership of farmland is appropriate and even desirable to the extent that it facilitates high levels of agricultural productivity. While those who can claim the farmer identity have staked out the highest moral ground, rentier landlords who improve their properties, or even just keep them in production, are still on firm moral footing. According to this moral framework, the only truly illegitimate farmland acquisition is the one that is left idle (something that rarely happens in well-funded investments). This limited perspective—which we might call the "politics of productivity"—precludes scrutiny of the other issues raised by rentier ownership, such as the justifiability of collecting ground rent in the first place, the social impacts of concentrated landownership, or the idea that land should belong to those who work it.

There is nothing straightforward about the value of land; economists have been grappling for centuries with the question of how land's worth as a means of production interacts with its worth as a scarce asset. For the finance-sector investors involved in the recent land rush, this interaction takes a particular form; in accordance with the ascendant logic of the financialization era these landowners depend heavily on financial returns—particularly the capital gains from land price appreciation—to achieve their desired level of profitability. This pursuit of capital gains leads many to actively convert their farmland to higher-value and more intensive uses, including clearing forested land and increasing yields on existing farm properties. The importance of capital gains to investment decision making can also be seen in the future-oriented narratives farmland fund managers and industry consultants deploy to attract investors; every investment pitch and market report conjures investor expectations and, with them, land price appreciation. Yet despite this fundamental focus on financial returns, farmland investment promoters work hard to position themselves among the producers of the world. This allows them to partially sidestep the long-standing stigma against those who earn rentier income and to bask instead in the moral glow associated with farming.

However, the *moral* sanctions surrounding landownership are not the only obstacle to farmland financialization. The farmland investment industry also encounters *material* and *political* hindrances as it works to transform land into something more closely resembling a standard financial asset class.

# MATERIAL DIFFICULTIES

**You know, farming looks mighty easy when your plow is
a pencil, and you're a thousand miles from the corn field.**

—Dwight D. Eisenhower, address at Bradley University, Peoria,
Illinois, September 25, 1956

Financial landowners may treat farmland as a financial asset—as a flow of monthly
rental payments and a swell of annual appreciation on their books—but it is not
one. In reality, the smooth circulation of capital through global farmland markets is greatly complicated by the unusual physical and social qualities of farmland. The challenge for farmland investors is to profit from land's positive aspects
without being ensnared by its dangers. This means overcoming the inconvenient
materiality of land and agriculture. Investors want land to behave more like a
stock or bond and less like an unwieldy expanse of dirt—while at the same time
extracting the benefits that stem from its expansive dirtiness. In pursuit of this
goal, those promoting farmland investment have debuted investment strategies
that minimize risk and increase flexibility, as well as investment metrics that make
farmland legible to a wider investor audience. Together, these approaches serve
to incrementally advance land's commodity status, smoothing the circulation of
capital through farmland markets despite land's awkward physical and temporal
characteristics. However, their success is far from being a foregone conclusion, and
financialization introduces new types of risk into the equation.[1]

## Natural Barriers, Financial Detours

The material qualities of farmland are a big part of its draw. On the plus side,
land is productive, scarce, and solid to the point of indestructibility. In stark and

pleasing contrast to the triple-A-rated mortgage-backed securities that went up in smoke during the 2008 financial crisis, land boasts unparalleled durability. What's more, because it is scarce, its value increases over time, something that can be said of very few other commodities. However, the unique materiality of land also creates problems for the circulation of capital: land is immobile, it is vast, it is expensive, it is heterogeneous, its markets are relatively illiquid, and in many places landownership is fragmented.

Just as land makes for a reluctant commodity, agriculture has characteristics that forestall its complete conversion to capitalist production. Marx and his followers were perplexed by the fact that capitalism, which had swept through urban areas like a factory fire, wasn't making the same inroads into the countryside. Capitalism, it seemed, merely smoldered around the edges of agriculture, producing smoke but few flames. Karl Kautsky threw himself into the investigation of what he called "the agrarian question": "*whether, and how, capital is seizing hold of agriculture, revolutionizing it, making old forms of production and property untenable and creating the necessity for new ones.*"[2] He answers this question largely by arguing that the need to work with land creates speed bumps in the capitalist conversion of agriculture.[3] In the twentieth century, this idea was elaborated on by Susan Mann and James Dickinson, Margaret FitzSimmons, David Goodman, and others, who argued that agriculture confronts capitalists with a phalanx of material and temporal obstacles, each presenting its own logistical challenge or risk: the seasonal labor requirements, the production process spread over many acres, the need to work within the limits of plant and animal biology, the unpredictability of weather and pests, and the perishability of the finished commodity.[4] In short, nature makes for a very temperamental business partner.

Yet it is also important not to overstate the obstacles that agriculture throws in the way of capital.[5] David Goodman and colleagues observe that industrial capital has managed to reduce many of the spatial and temporal difficulties presented by nature through technological innovations at discrete stages in the production process; fast-moving tractors, high-yielding crop varieties, and confined-animal agriculture all reduce the spatial constraints of land-based production, while the use of biotechnology instead of traditional plant and animal breeding and the development of quick-maturing crop varieties reduces the time constraints implicit in biological reproduction.[6] Patrick Mooney, meanwhile, describes a series of "detours" through which capitalism has subtly penetrated family farming, including contract work, hired workers, off-farm employment, tenancy, and debt. The capitalist transformation of agriculture, he argues, is taking place in many complex ways that extend beyond the direct act of farming. Can we still speak of the autonomous family farmer, for instance, if a landlord owns his land and the bank owns his machinery?[7]

The financial sector has been particularly resourceful at fostering capital circulation through agriculture. George Henderson argues that the very same obstacles that nature places in the path of *industrial capital* may create profit-making opportunities for *financial capital*. Those same temporal discontinuities and uncertainties that make agricultural production problematic for producers, he contends, create lucrative openings for moneylenders and speculators. Farmers are perpetually in need of cash to bridge the several-month gap between planting (when large amounts of money are needed to pay for inputs and labor) and harvest (when the farmer finally has a product to sell). Add to this the enormous capital needed to purchase land in the first place, and it becomes clear why farmers are so often in debt. Lenders—including banks, life insurance companies, and input providers—help farmers bridge the gap between present expenses and future income, making a tidy profit in the process.[8] Through finance, capitalism can infiltrate agriculture without ever getting its (invisible) hands dirty.[9]

Today's farmland investment intermediaries have introduced new strategies for skirting or even profiting from land's intransigence. Adopting Mooney's language, we can think of these investment strategies as "detours" around the long-standing material barriers to profiting from farmland. Six financial detours are considered below: (1) portfolio diversification as a detour around agricultural risk, (2) commensurative metrics as a detour around land's uniqueness and heterogeneity, (3) sophisticated uses of digital data as a detour around land's expansiveness, (4) private-equity-led consolidation as a detour around land fragmentation, (5) operating companies separating landownership from production as a detour around land's high cost and nondepreciability, and (6) publicly listed farmland real estate investment trusts (REITs) as a detour around land's illiquidity. The first three involve conceptual advances in understanding, measuring, and visualizing farmland investment, while the latter three involve the creation of new investment vehicles for actually acquiring land. None of these solutions are without historical antecedents, but all have been taken to new levels of sophistication in recent years. As Henderson puts it, "Capital and nature are webs of constraint and confinement that must be carefully recast as fields of opportunity."[10] Over the past decade, the farmland investment community has been busily at work forging opportunity from constraint.

## Diversifying Away Risk

One of the biggest difficulties investors face is the riskiness of agriculture.[11] At the time of planting, it is impossible to know for certain whether drought or flood or frost or hail (or mite or blight or rust, for that matter) will obliterate your investment several months hence. Alternatively, good weather in too many places can lead to oversupply and falling prices. Furthermore, once successfully harvested, perishable crops must be consumed within a given time frame and can spoil if not handled properly. Goodman and colleagues suggest that the central constraint confronting capital in its attempts to penetrate agriculture is "the inability to eliminate the risks, uncertainties, and discontinuities intrinsic to a natural or biological production process."[12] Risks vary by crop; the major risks associated with lettuce production, for instance, lead Friedland et al. to observe that "lettuce is a more speculative commodity than most other agricultural products" and that lettuce growers have a reputation for being gamblers.[13] However, all agriculture abounds with risk, which basically comes down to socio-natural processes making the future unpredictable.[14]

To cope with these risks, farmland investors today deploy a deceptively simple strategy: diversification.[15] In recent decades, diversification has emerged as one of the star players in finance economists' efforts to deal with future risks. The US economist Harry Markowitz pioneered an approach to investing known as modern portfolio theory (MPT), which essentially makes a quantitative case for not putting all your eggs in one basket. According to this theory, investors should not focus on choosing individual stocks with the highest possible return or lowest possible risk but should instead focus on how all the investments in the portfolio work in combination. An optimal portfolio will contain investments whose risks are unrelated to one another so that they over- and underperform under different conditions, ensuring that the portfolio continues to make money no matter what. In other words, under the influence of MPT, investing becomes less about picking a winner and more about fielding a good team.[16] By carefully selecting the exact right mixture of assets, an investor can create a portfolio at the "efficient frontier," the term used to describe portfolios with the lowest possible level of risk for the desired level of return, or the highest possible return for the desired level of risk. Though investors widely embraced the financial benefits of diversification in the 1970s, the idea of applying it to farmland has caught on much more recently.[17]

This growing theoretical understanding of diversification has carved a detour around agricultural risk for two reasons. First, as finance became the science of curating a perfectly balanced portfolio, the idiosyncratic risks associated with land and agriculture transformed from deal breaker to attraction. It is hard to overstate the importance of this shift for the field of agricultural investing: suddenly, *unusual risks became useful* as a means of reducing the vulnerability of a portfolio to overall market fluctuations. The drive to diversify contributed to making real estate as a whole more attractive to investors because real estate is not highly correlated with stocks and bonds. The push for portfolio diversification was one of the catalysts behind institutional-investor demand for US timberland throughout the 1980s and 1990s. When the food and financial crises coincided in 2008, it was farmland's moment to be thrust into the spotlight as a possible ingredient in the much-sought-after "risk-efficient" portfolio. For the owners of capital, the weird risks of agriculture suddenly started to look like a blessing rather than a curse.

Second, diversification *within* a portfolio's farmland allocation can serve to further reduce risk. This idea has only recently come of age. A 1985 article in the *Journal of Portfolio Management* floated the possibility as a hypothetical: "Investors are averse to farming return volatility. Nevertheless, if the farm operation reached a large enough scale, the investor could begin to diversify away the volatility."[18] Or, in our terms: agricultural risk is a barrier to investment, but diversification offers a detour. By the 2000s, the idea of the diversified farmland portfolio had taken off, becoming something of a mantra among investors. One operating-company executive I interviewed, for instance, described four different axes along which his company diversifies: "We are very big on diversification. We diversify *by crop*—that means you have different planting and harvesting periods; *by region*, so that the same weather event doesn't affect you in the same place; *by political domicile*, so the same stupid set of politicians can't screw you everywhere; and diversify *by going vertical* where it makes sense, and then you can reduce the variability of your top line."[19] Though vertical integration was relatively unusual among the investors I interviewed (many of whom owned only the land), the other three forms of diversification were very common. By diversifying their holdings, they reduce their vulnerability to the many natural and social perils that populate the possible future. In short, capital no longer needs to civilize nature, just counteract it with opposing tendencies.

Of course, investors did not invent the idea of diversifying to reduce agricultural risk. Traditional agricultural systems depend on diversification—employing crop rotation (alternating the crops planted in a given field), intercropping (planting diverse crops in the same field at the same time), or both. These strategies serve as a means to reduce weather risks, maintain soil health, and smooth labor demand throughout the year.[20] But here we see diversification on a completely different scale and level of sophistication. The mobility of financial capital and the sheer quantities involved allow for cross-sector, cross-continental diversification to a degree not seen before. And unfortunately for the environment, a globally diversified farmland investment portfolio can reduce agricultural risk without abandoning industrial mono-cropping.

Geographic diversification brings its own kinds of risks, though these too can be mitigated to some degree with the right investment strategy. Fluctuating exchange rates are a major source of risk in international investments, but currency hedging can reduce that risk. An asset manager I chatted with at an agricultural investment conference cocktail hour explained this risk by way of a detailed hypothetical example. Suppose, he said, that your fund has $10 million that it plans to spend on Australian farmland. At present the exchange rate between US dollars and Australian ("Aussie") dollars is 1:1, but you believe that the Aussie dollar is overvalued and is therefore likely to drop in value in future. This could wreak havoc with your investment, since the land must be purchased in Aussie dollars. It would be very unfortunate if the value of your Australian farmland appreciated from $10 million to $11 million US, but in the meantime the Aussie dollar lost so much value relative to the US dollar that you were nonetheless left with only $9 million US when you sold the land at the end of the investment period. You can avoid this situation by using forward exchange contracts or similar foreign-currency transactions to guarantee yourself a set exchange rate at a future date. Exchange-rate fluctuation is a socially manufactured risk, but it is the physical immobility and irreproducibility of farmland that necessitates purchasing it in foreign locations and currencies in the first place. If investors could just build more farmland at home, then currency hedging would be unnecessary.

International farmland investors can also take advantage of political risk insurance, such as that provided by the Multilateral Investment Guarantee Agency (MIGA) of the World Bank or the US's Overseas Private Investment Corporation (OPIC). In recent years, MIGA has provided tens of millions of dollars in political risk insurance to support agricultural private equity companies investing in sub-Saharan Africa, including Chayton Atlas Agricultural Company and SilverStreet Partners.[21]

## Addressing Land's Uniqueness and Its Heterogeneity

The material qualities of farmland also present a suite of problems relating to financial presentation and legibility. In particular, before land can become a financial asset class, promoters must tackle two issues: first, that land is *unique*—it is fundamentally unlike other commodities both physically and socially; and, second, that land is *heterogeneous*—two plots of land in different parts of the world will be quite unlike one another. In other words, land exhibits a confounding singularity and an equally confounding multiplicity. Both must be addressed before it can become a truly useful conduit for investment capital.

The nascent farmland investment sector has tackled these problems through various acts of commensuration—the process of making qualitatively different things comparable to one another via quantitative metrics.[22] To address land's uniqueness, farmland investment promoters facilitate comparisons with other asset classes, often via metrics of risk and return. Figure 3.1, published by the asset management firm GMO, shows how US farmland, when reduced to percentage points of risk and return, is rendered perfectly commensurate with bonds, treasury bills, and US small cap equities. Two numbers subsume the materiality of

**Historical Risk & Return for U.S. Timberland & Farmland and Selected Asset Classes**
January 1994 - December 2013

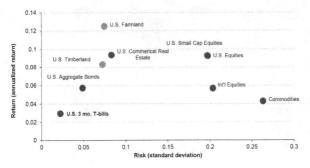

Source: Morningstar, NCREIF
See endnote for information regarding indexes used in this exhibit

**FIGURE 3.1**    Figure from a 2014 GMO white paper comparing farmland to other asset classes
*Source:* Koeninger, "Farmland Investment Primer."

soil, worms, bees, weeds, rain, frost, fertilizers, and tractors, as well as all social relationships with tenants, neighbors, and laborers, allowing farmland to join its fellow data points for the ready appraisal of any interested investor. Such numerical representations of land are linked to the unique mentality of financial landowners discussed in chapter 2: though family farmers may closely track the current market value of their farms and make decisions based partially on speculative criteria, they are unlikely to sell their land purely because its risk-return profile has declined relative to commercial real estate or international equities; financial landowners, on the other hand, are likely to do exactly this.

Another problem with farmland as an asset class is its total lack of uniformity. As the investor gaze wanders over the globe, it encounters a bewildering array of soil types, climates, land-tenure arrangements, and political regimes. Attracting nonspecialist investors to the sector requires, therefore, that they be given some basis for comparing these disparate properties.

There are many possible metrics for commensurating farmland across the globe, and the nascent farmland investment industry is busy creating new ones. The London-based global real estate company Savills is at the forefront of this effort. In 2012 it introduced a "Global Farmland Index," which converts farmland prices from fifteen countries into US dollars and expresses them as an index to allow for easy comparison between countries and regions.[23] Figure 3.2 shows the index in action; it allows for an almost instantaneous comparison between rates of farmland appreciation in different global regions. This rendering is elegantly simple, but also strikingly low resolution, reducing vast geographic and political complexity into a single dollar value per country. But reducing complexity is, after all, the entire point of commensuration.

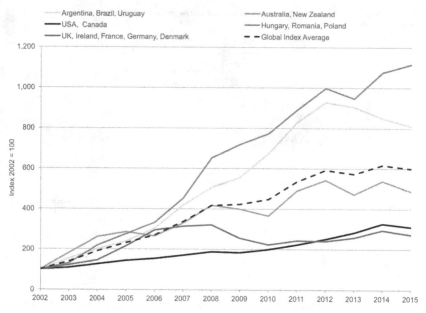

**FIGURE 3.2**   Savills Global Farmland Index, 2017
*Source:* Savills Research, based on data from various sources.

Savills makes a more advanced bid for comparability with its farmland "Opportunity v. Risk Matrix Tool," in which countries are assigned a risk and a benefit score, allowing for cost-benefit analysis. According to Savills, the risk score given to each country incorporates a mind-bogglingly diverse "matrix" of factors, including political risk, currency volatility, climate, fiscal policy, GDP growth, liquidity of the farmland market, infrastructure, subsidization of agriculture, research and development activity, and agronomic potential. The benefit score incorporates scalability, soil quality, potential for yield/output increases, water availability, and sustainability. While some of these categories (such as GDP growth) are relatively straightforward, others (such as agronomic potential and sustainability) are more mysterious.[24] Figure 3.3 displays a recent iteration of the matrix, ranking countries in consecutive order based on the overall "score balance" between costs and benefits.

It is easy to forget that such numbers are social artifacts. Though designed to appear self-explanatory, intelligible to even the most casual evaluative glance, they are in fact the product of complex human calculations and decisions. They contain their own kind of narrative based on particular decisions about what is worth counting.[25] Consider, for instance, the NCREIF Farmland Property Index, referenced in the previous chapters. Promoters of farmland investment frequently use the NCREIF index to show prospective investors the kind of returns

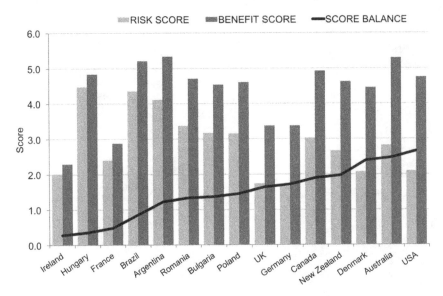

**FIGURE 3.3**  Savills Opportunity v. Risk Matrix Tool results, 2017
*Source:* Savills Research, based on data from various sources.

they can expect from US farmland. In doing so, they are unlikely to mention exactly how the index is calculated, yet this information is vitally important. For one thing, NCREIF takes into account only farm-level management fees, but not the portfolio-level fees that many farmland investors pay. For another, the set of farm properties about which NCREIF collects data includes a higher proportion of permanent crop properties than the US average, which matters because permanent crops generally produce more income than annual crops. Finally, the index excludes farms that are not yet fully operational, meaning that the first few, relatively unprofitable years in which new infrastructure is being installed or newly planted permanent crops are reaching maturity are left out of the calculation.[26] These decisions have the effect of boosting the farmland investment returns the NCREIF index reports.[27] (It is tempting to say that the numbers are "inflated" or "overstated," but this presumes that a "true" farmland return figure exists out there somewhere, independent of human efforts to calculate it. It does not.) Yet such metrics are rarely subjected to critical scrutiny; they appear as unimpeachable fact, and in this lies their power. The NCREIF index has been such an important tool in the effort to turn US farmland into a standardized asset class that in 2016 the NCREIF methodology was used as the basis for a new Australian Farmland Index.[28]

Private-sector actors aren't the only ones working to make farmland thinkable as a global asset class. Figure 3.4 displays a widely cited graphic from the

Figure I    Potential Land Availability vs. Potential for Increasing Yields

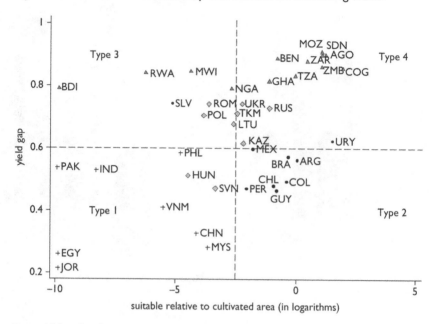

Source: Authors based on Fischer and Shah 2010.

**FIGURE 3.4**    World Bank graphic: Potential land availability vs. potential for increasing yields
Source: Deininger and Byerlee, *Rising Global Interest in Farmland,* based on Fischer and Shah, "Farmland Investments and Food Security." Available under World Bank CC-BY license.

2011 World Bank report *Rising Global Interest in Farmland.*[29] In this figure, the report's authors classify all global farmland into one of four groups based on two variables: the x-axis represents the amount of sparsely populated, uncultivated land a country has "available," while the y-axis represents the "yield gap" between a country's current and potential agricultural productivity. Based on these two metrics, the authors explain that the greatest investment opportunities lie in countries in the upper right-hand corner of the graph. These "Type 4" countries are deemed to have a lot of land that is not in production and to be underutilizing what land they do produce on, making them perfect targets for foreign direct investments in agriculture. The World Bank's legitimacy as a development institution derives largely from its reputation as a source of ostensibly neutral data,[30] but such commensurative metrics include implicit value judgments. The very concept of "underutilized land" reduces the quality of an agricultural system to the extent of its yields and proclaims that more is always better. At a farmland investment conference in 2013, I saw this image displayed to an audience of prospective investors as a representation of global investment opportunities; in this moment, the World Bank's seemingly technical assessment was effectively deployed as a means to further the financialization of farmland.

## Monitoring Vast Amounts of Land with Digital Data

The heterogeneity of land, combined with its physical expansiveness, also presents difficulties *after* investment capital has been raised. Acquiring and efficiently managing hundreds or thousands of acres of land, often geographically dispersed between states or countries, is a colossal logistical challenge. Though this is a rapidly developing area, it appears likely that the farmland investment industry will address these challenges with data. Digital data has been transforming agriculture since the rise of "precision agriculture" in the 1990s, which uses geographically referenced data to guide farm machinery and tailor inputs to the needs of particular sites within fields.[31] The early 2010s saw the further development of "digital agriculture" as advances in sensor technology, big-data analytics, and cloud-based storage allowed for more sophisticated uses of satellite and other remotely sensed data.[32] By now, digital data has become virtually indispensable to large-scale farm management, a fact confirmed by a spate of data-driven mergers and acquisitions within agribusiness, including Monsanto's 2013 acquisition of weather-data startup Climate Corporation for a whopping $930 million.[33] Digitalization facilitates large-scale agricultural production by allowing operators to visualize and understand expanses of land too big for them to regularly visit and experience in person.[34]

Farmland investors, who are often managing many properties at great distances, are another logical clientele for data-driven farm management services, a fact that companies are beginning to realize. CiBO Technologies is an example of one such company. Founded in 2015, CiBO uses big-data analytics, statistical modeling, and artificial intelligence to assess farmland value.[35] At a recent agricultural investment conference, a CiBO representative explained that there are "literally hundreds of thousands" of data points to consider when it comes to land. These include its location, previous crops, water sources, organic or other certifications, proximity to interstates and airports, proximity to refrigeration, and much more. Whereas farmers with just a few properties can keep track of these variables easily, really large growers as well as institutional investors with huge portfolios of farmland require a much more sophisticated data management system. The CiBO presenter told the audience, "If you think about old school family farming, so four or five generations ago, you had a handful of parcels ... you could manage it on a napkin. ... When you get to tens to hundreds of parcels you move into the Excel spreadsheet. . . . But then when you get to hundreds to thousands of fields . . . sometimes spread across the entire country, sometimes even international, you now need something like artificial intelligence and machine learning." Whereas the physical extensiveness of agricultural land might formerly have limited the number of farms that investors could amass and manage profitably, advances in data collection, storage, and analytics promise to remove the upward limit on landownership consolidation.

Another company offering digital data services to farmland investors is Tillable, a farm data management platform launched in 2019 that focuses on mediating the landlord-tenant relationship. Tillable allows investors to comparison shop between prospective farmer tenants, ensuring that investors receive the highest possible rental income for their properties. It also automates collection of data from tenant farmers and allows investors to monitor tenant performance and compare it to benchmarks.[36] Whereas the need to oversee and interact with many tenants might formerly have been a barrier to rentier landownership on too vast a scale, digital data promises to facilitate monitoring—one might even say surveillance—of tenant farmers, transforming them into "data subjects" whose practices are increasingly legible to (and controllable by) their landlords.[37]

## Assembling the Private Equity Jigsaw Puzzle

A major spatial difficulty that capitalist producers encounter is the need to assemble large tracts of farmland from smaller ones. Karl Kautsky observed that while in most industries producers grow by first expanding their business and then gradually putting smaller competitors out of business, in agriculture it is necessary to proceed in the opposite order: agricultural producers must expand by purchasing the properties of neighboring smallholders. Where landownership is fragmented and private property prevails, this work of assembling a large farm is a painstaking process, particularly as the small farms assembled should be contiguous, and individual smallholders may refuse to sell. In short, capitalist agriculture can come to resemble a particularly annoying jigsaw puzzle—one in which the pieces are fiddly and some have gone missing from the box.[38]

Yet while frustrating for agribusinesses, land fragmentation provides an opening for the branch of finance known as private equity. Private equity investment involves buying up companies and restructuring them to generate greater investor profits; this could mean breaking companies up and selling off the parts or fostering mergers between companies. Private equity, in short, specializes in assembling and disassembling productive assets. The slew of new farmland funds that began to emerge in the mid-oughts included many funds with a private-equity-like structure. These funds either acquire farmland-owning agribusinesses or invest directly in a portfolio of properties. Like typical private equity funds, many are set up as limited partnerships in which the asset manager is the "general partner" and the investors are the "limited partners." Many also use a fee structure common to private equity and hedge funds in which investors pay the asset manager an annual management fee (generally around 2 percent of their committed capital) as well as a performance fee or "carried interest" (generally around 20 percent) once the fund is successfully wrapped up.[39] Some asset managers also engage in farmland equity co-investments, in which loyal institutional clients are given the opportunity to invest alongside the private equity managers for significantly reduced fees.[40]

Farmland funds modeled on private equity operate on a fixed term, often of seven or ten years, and the ultimate goal is always a successful "exit." Only by exiting the fund can investors receive their return on capital and managers their performance fee. The most common exit strategies are selling off the properties to a strategic buyer, such as an agricultural operating company; taking the entire fund public via an initial public offering (IPO) on the stock market; or rolling the properties over into a new fund. This last option allows some investors to remain in the fund, keeping their stake in the farmland, while others have a chance to get out and new investors can come in. The significance of this time frame is entirely relative. For investors accustomed to the frenetic tempo of the stock market, with its daily index performance and quarterly earnings reports, private equity represents a very long-term investment. However, in the context of agriculture, where farms often stay in the same family for generations, a decade between purchase and resale represents a shrinking time horizon.

Like all private equity funds, these farmland funds have an obligation to produce shareholder value. A major selling point is their ability to generate "alpha" returns—those returns that stem from managerial prowess above and beyond the so-called "beta" return generated by baseline market movements. Funds that acquire entire agribusinesses can generate shareholder value through typical private equity strategies—streamlining operations, selling off parts of the company, and downsizing employees—but for funds that primarily buy land, a different set of strategies is needed. Shareholder value could come from the skilled identification of undervalued land that is likely to appreciate (basically farmland arbitrage), or from improvements to farm soil or infrastructure. Land fragmentation may also offer an opportunity to create shareholder value. In regions where landownership is highly fragmented among smallholders, farmland private equity funds are buying many small properties with the express intention of consolidating them into large farms that will be candidates for eventual sale to agricultural operating companies. This strategy is particularly common in Eastern European and former Soviet countries where collectivized farmland was redistributed after the fall of communism, leading to a landscape of very small landholdings.

In the spring of 2012, for instance, I interviewed the manager of a fledgling Eastern European farmland fund, a wiry man of indeterminate age who was still working his day job at a brokerage firm in lower Manhattan. We met in his shared office space, his coworker slipping on headphones as we began to talk. He told me that because land had been redistributed and then further divided through the generations, plots in Bulgaria and Romania are too small to be useful—"little tiny slivers of land" as small as ten acres, he said. "It's ridiculous." His fund identifies the owners of these fragmented plots and buys them at market price, which he emphasized was cheap both in the context of Europe and in comparison to what similar land would cost in plots of a larger size. They then consolidate the plots into larger properties as a means to raise their value. Consolidation, he said, leads to instant appreciation: "A plot can be worth more money immediately by 25, 30, even 50 percent more if you just merge three or four together." Thus, the very same fragmented ownership structure that hinders capitalist producers constitutes a source of potentially quite impressive returns for his fund.

Consolidation may generate impressive appreciation, but it is not easily accomplished. Two weeks later, I interviewed a manager of a well-established fund whose strategy also hinges on consolidating the post-communist landscape, particularly in Russia and Ukraine. This time we were in an opulent high-rise in Midtown, seated in a conference room with a majestic view over the Manhattan cityscape. The manager described how his company goes about assembling the farmland jigsaw puzzle:

> You've got to go identify the property. You have to identify it in a region that is fertile and largely fallow. You have to then understand if there are any people who are going to hold out and that kind of thing. And you start building in the center and you build out toward the edges and you keep going until you think you've got the appropriate size. In Russia, very often the people who are living on that 2.7-hectare farm don't even understand what they have. They are not educated people; they have a piece of paper that says they own the land. Well, when we look at the piece of paper, we realize that that's not what it really says. It doesn't have, you know, the fourteen stamps, the three seals, the eight signatures, and that kind of thing. So what has taken us the most time is having people actually go out and help that farmer register his or her own land so that then they can sell it to us, which we have to go re-register. . . . There is the proverbial sixty-three thousand pages of documents that are needed just to get one farm put together.

In order to consolidate small properties, his fund must first shepherd them into the formal market, taking them through the final, official steps to privatization. His company is therefore adding value simultaneously through formalization of title and consolidation of small properties.

When I asked him what the "appropriate" size was for these consolidated farms, he told me that their target was 10,000 ha (which is close to 40 square miles). I asked why that size, and he emphasized the economies of scale involved in large-scale production: "I'm not in the field doing this, but my understanding is 10,000 hectares is the point at which you have the right number of people, the right number of machines, they are optimally being used, and you are making money off enough so that it counts. . . . You got to have scale." His explanation echoes Kautsky's argument that scale is essential to capitalist agriculture. Whereas the difficulty of assembling such large-scale farms might pose a barrier for large agricultural operators, it also provides a potentially lucrative opening for private equity funds.[41] His explanation also suggests a detachment from the irksome ins and outs of material engagement with agriculture—he does not personally know the details of farm operation, but he knows what size farm will sell, and this is the point upon which his profit depends.

## Unlocking Value from Operating Company Properties

In the financialization era, corporations are constantly scrutinized by shareholders for signs of good or bad performance. Company financial statements and reports by market analysts provide fodder for shareholder decisions about whether to buy or sell stock in a company. Unfortunately, however, land does not always play nicely with the other numbers.[42] The high cost, nondepreciability, and disparate risks involved in farmland ownership all conspire to make nonfinancial companies that own farmland—agricultural operating firms primarily—look bad on paper.

One major bookkeeping problem encountered by landowning companies relates to the high cost of land. Performance indicators such as return on assets and return on capital put the company's annual earnings in the context of all the resources the company potentially has to work with. Because land is so expensive, farmland-owning agribusinesses may appear not to be getting the most out of their assets. This perceived inefficiency is damaging at a time when the shareholder-value view of corporate governance creates pressure for companies to slim down their operations, whether that means downsizing the workforce or selling off physical assets.[43]

A second source of financial-statement headaches for landowning companies is, paradoxically, a side effect of one of land's most attractive qualities: its inevitable appreciation. Most corporate assets depreciate over time, which is to say that they become less valuable as a result of the wear and tear of daily use. Buildings and equipment depreciate, as does living agricultural "infrastructure" such as trees, grapevines, and even livestock. Farmland, however, does not; its price fluctuates, but is assumed always to increase in the long term. This is a problem because the rules governing the creation of public companies' financial reports—such as the Generally Accepted Accounting Principles in the US—usually allow companies to declare the initial capital outlay for depreciable assets as a tax-deductible expense over the years that follow.

This issue was laid out for me by Robert, an avuncular senior executive at a pension fund with extensive farmland investments. Over bagels in Manhattan, Robert explained simply that "it juices your returns when you depreciate." Public companies with a large amount of their fixed assets in farmland are therefore put at a disadvantage, relative to other public companies, by this inability to depreciate. Robert summed it up like this: "If you are investing in assets that don't maximize your balance sheet and your income statement, then you are not investing efficiently, and I can invest better elsewhere. And so stocks become undervalued if they have any kind of asset which is nondepreciable. That's just a fact of life."

A third difficulty encountered by the farmland-owning agribusiness is the differing types of risk involved in landownership and agricultural production. Such companies are not a "pure play"; they represent an investment in farmland but also an investment in agricultural production, and perhaps food or fuel processing as well, which not all investors want. In interviews, several investors discussed this as a possible reason why the stock of public operating companies often underperforms. As one farmland investor put it,

> Markets perceive that there is a different risk and there is a different return in owning land than there is from operations on the land. I mean, it's hard to make a piece of land disappear, okay. So, you know, land in itself is a fairly risk-free asset to own.... But the operations on top of that land have more risk. So I think that when they look at operations, they say, "Well, there could be drought, there could be hail, there could be spikes in the value of corn or wheat." Farming is a whole different business.

Stock in such companies, he argued, tends to be undervalued as a result of this investor confusion about how to value this mixture of dissimilar risks.

With growing investor interest in farmland during the recent boom, some large agricultural operating companies saw a possible detour around these problems: "spinning off" part of their farmland portfolio into a separate asset management business. They are not the first property-owning businesses to hit on this solution. Mounting pressure to increase shareholder value over recent decades has led many companies to remove expensive assets from their books by selling or leasing them out. This was largely what prompted vertically integrated US timber companies to sell off their land to institutional investors during the 1980s and 1990s (see chapter 1).[44] Where farmland is concerned, this strategy is most common in countries where a concentrated landownership structure has already made it possible for operating companies to own hundreds of thousands of hectares of land. Brazil is one such country.

The case of the publicly traded Brazilian agribusiness SLC Agrícola is illustrative of this approach. In 2012, this cotton, corn, and soy producer announced the creation of a separate agricultural property company called SLC LandCo. In order to construct LandCo, SLC took 61,000 ha of its existing 200,000-ha cropland portfolio and used it to raise about US$240 million from the British asset-management firm Valiance in exchange for a share of just under 50 percent in LandCo.[45] These funds are being used to purchase additional agricultural land with potential for rapid appreciation, all of which will continue to be operated by SLC Agrícola.

Another public Brazilian agribusiness, the sugar-alcohol-sector company Cosan, adopted a similar approach. Cosan is one of the largest companies in Brazil, a conglomerate that owns hundreds of thousands of hectares of land in addition to many sugar mills, ethanol refineries, and other assets. In 2008, Cosan collaborated with TIAA to create Radar Propriedades Agrícolas, a rural real estate business. As the Cosan website explained, Radar aims to "capitalize on new business opportunities in the Brazilian rural real estate market, purchasing properties with significant potential for appreciation and leasing them to major agricultural producers. After they reach their target value, the properties are put on the market."[46] With the Radar venture, Cosan's prowess as a rural real estate player was isolated from its agricultural activities, providing a more attractive investment target for its pension fund investor.

Operating companies' move to place some of the land they already own in a new, farmland-focused management company reflects a broader trend toward viewing real estate as a source of unrealized financial value. This trend can be seen in the "opco/propco" restructurings undertaken by some supermarkets and other property-owning businesses. In this approach, the company divides into two entities: an operating company (opco) that runs the business and a property company (propco) that manages the property upon which the business sits. Both entities have the same owners, and the operating company continues to use the property by leasing it from the newly established property company. These restructurings rest on the idea that property contains pent-up value that can only be released through the right institutional arrangements.[47] As a 2010 report by agricultural consulting firm HighQuest Partners explains, "The rationale for this move is twofold: to unlock 'hidden value' in [the company's] equity which trades at a discount to its net asset value and to create a platform for raising capital from a larger universe of investors which maintains a preference for land ownership (a hard asset) over investing in farm management operations." Value is "unlocked" insofar as new investors are attracted to the project by the opportunity to get exposure to the real estate side of the business alone.

The creation of farmland real estate companies and funds from within the agribusiness sector is emblematic of nonfinancial firms' growing reliance on financial profits under financialization.[48] As discussed in chapter 2, farmland (like much real estate) occupies a strange middle realm between the financial and the productive—it produces productive income from farming but also financial income in the form of capital gains from appreciation. The move Cosan and SLC made is significant because it shows their reconceptualization of their farms, from substrate for agricultural production to rapidly appreciating financial asset that can be used to raise additional investor capital, from source of use value to source of exchange value. SLC Agrícola underscored this shift with a new slogan introduced in 2012, which has since graced the beginning of all its English-language investor presentations: "SLC Agrícola: Value from Both Farm and Land" (see figure 3.5). Land is no longer just a necessary agricultural input; it is a source of financial profits in its own right.

**FIGURE 3.5**   Introductory slide of SLC Agrícola investor presentation, 2016
Source: SLC Agrícola, "March 2016 Presentation for Investors."

# Increasing Land Liquidity with REITs

The idea of "unlocking" value points toward one of the biggest difficulties associated with farmland ownership: the way that it immobilizes capital. Land is expensive, and the low liquidity of land markets makes invested capital difficult to extract again. Because farmland markets are generally local, because real estate transactions are time consuming, and because many farmers have a tenacious attachment to their land, farms can take a long time to buy and sell. The cliché financial advice to "buy low, sell high" therefore assumes precisely what land-market investors often lack: a choice of timing. The stock-market investor who believes that a certain stock will soon increase in value can buy into it almost instantaneously, while the land-market investor in the same position may not be able to find a suitable plot of land to purchase for a matter of weeks or months. Likewise, if she subsequently needs the capital for another investment, a rapid sale may not be easily accomplished. She may therefore have to watch as energy or telecom or biotech stocks climb at dizzying speeds while a significant chunk of her capital is rendered infuriatingly inaccessible in its terrestrial vault. In addition, land markets are not very accommodating to investors of different sizes; the large investor may have to buy many small farms to absorb enough capital, incurring considerable transaction costs in the process, while the small investor may find the high up-front cost of buying even a single farm to be prohibitive.

Those concerned about liquidity have the option to forgo direct farmland ownership and instead buy stock in a publicly traded farmland-owning agribusiness, but this entails investing in agriculture as well. However, the recent farmland boom introduced a more complete detour around land's illiquidity: the public farmland real estate investment trust (REIT, pronounced "reet"). A REIT is a special type of company focused on acquiring income-producing real estate. It is exempt from paying corporate taxes because it distributes 90 percent of the income from this real estate directly to shareholders in the form of dividends.[49] When a REIT goes public, it issues securities that are traded on the stock exchange. This investment vehicle transforms farmland from a notoriously illiquid investment to a highly liquid one and allows regular, nonwealthy individuals—what the financial sector calls "retail investors"—to buy into farmland on a small scale. REIT stock values (and payments to shareholders) are based on farm rents, making the stockholder a sort of mini-landlord taking a bite-size cut of the rent.

Public farmland REITs first appeared in Bulgaria, where between 2005 and 2006 at least five were created with the view of profiting from inevitable land-price increases when Bulgaria joined the EU in 2007 and—like the private equity funds discussed above—from consolidation of the fragmented plots that resulted from post-communist land distribution.[50] The US has hosted public timberland

REITs for years and even vineyard REITs, but the January 2013 IPO of Gladstone Land Corporation marked the first time that stock market investors could buy into garden-variety farmland. Gladstone Land Corporation is a farmland-focused REIT based in Virginia whose parent company, Gladstone Investment Corporation, already ran a public REIT composed of commercial real estate. As of December 2016, Gladstone owned fifty-eight farms comprising 50,592 acres, valued at $401 million. It owns farms across seven states, with heavy concentrations in California and Florida. Initially focused on annual specialty crops like vegetables and berries, it has since diversified somewhat into permanent fruit and nut crops as well as some annual commodity-crop properties. The company takes no part in farm operation; its profits come from leasing the farm properties out to corporate and independent farmer tenants. It acquires land, in part, through sale-leaseback deals, in which the farmer sells land to the company in return for a long-term lease to continue as the farm operator.[51] Appropriately enough, the company nabbed the Nasdaq ticker symbol LAND—so investors can now literally buy stock in LAND.

Two other public farmland REITs have since debuted in the US: Farmland Partners in 2014 and American Farmland Company in 2015, both traded on the New York Stock Exchange. In 2016 these two companies merged under the Farmland Partners name. As of December 2016, the resulting company owned 152,000 acres of farmland valued at almost a billion dollars. Its farmland portfolio is distributed across seventeen states, with heavy concentrations in the Midwest, Southeast, and California. Compared to Gladstone, it is much more focused on commodity row crops (75 percent of its portfolio) and much less on specialty row crops and permanent crops (the remaining 25 percent).[52] In Australia, meanwhile, farmland REIT Rural Funds Group went public in 2014. By December 2016 it had a portfolio valued at $552 million, much of it invested in almond orchards and poultry sheds, but also including vineyards, macadamia nuts, cotton, and cattle.[53]

Public farmland REITs are not the only unusual financial vehicles aimed at making farmland accessible to retail investors. In the US, several companies are using "investment crowdfunding" to sell shares in farms on the internet. Crowdfunding is best known as a way to raise donations on the internet in support of artists or charities. However, the 2012 passage of the US Jumpstart Our Business Startups (JOBS) Act reduced the securities regulations that apply to crowdfunding. While crowdfunded companies could previously only compensate their "investors" with gifts like T-shirts and signed books, investors can now receive company debt or equity in return for their investment.[54] Hard on the heels of the JOBS Act came the first crowdfunding investment company dedicated to farmland. Fquare, established in 2012, allowed accredited investors to buy ownership stakes in operational Corn Belt farms for periods of three, five, or seven years in return for quarterly dividends drawn from the farm lease payments.[55] Fquare went out of business, but successors with similar models soon followed, including California-based FarmFundr in 2015 and Arkansas-based AcreTrader in 2019. Farmland crowdfunding platforms lack the liquidity of public farmland REITs, although AcreTrader has plans to create a secondary market that would allow its investors to buy and sell farmland shares among themselves.[56] These vehicles are similar to farmland REITs, however, in that they facilitate equity investments in land in increments much smaller than the cost of an entire farm. The Fquare home page framed this approach as "Fractional farmland owner-ship," continuing, "Mitigate the risk of farmland ownership by purchasing a share of a farm rather than the entire farm" (figure 3.6).

Though still in their infancy, public farmland REITs (and similar vehicles) have the potential to transform the meaning of farmland ownership radically. By turning farmland into a publicly traded security, public farmland REITs attempt to dismantle the barriers posed by farmland's high cost and low liquid-ity. They use securitization to take the stagnant capital embodied in farmland and set it in motion.[57] In this new form, farmland is cheap—its imposing up-front costs are replaced by a share price that anyone can afford with just a few bucks. It is liquid—shares can be bought and sold and bought again in a matter of moments. It is also fungible—an acre of land is one of a kind, but a share in a

**FIGURE 3.6**  Fquare website home page, 2014
*Source:* Fquare, "Home."

farmland REIT is thoroughly interchangeable. Such vehicles offer the American dream of landownership in a way calculated to make the modern portfolio investor feel at ease.

Another way to think about the farmland REIT is in terms of Marx's discussion of capital and commodity circulation. Marx argued that under precapitalist economic systems money served simply as a means of exchange between different commodities (he dubbed this capital circuit C-M-C). Under capitalism, however, production and exchange of commodities is primarily a means to convert money into more money (a circuit he termed M-C-M'). In financial transactions, the commodity becomes even less essential to the process, and money is put to work directly to making more money (M-M'), as is the case, for instance, with an interest-bearing loan.[58] Though investors who buy land through a separately managed account or commingled fund may already experience farmland as a financial investment to a very great extent, the public farmland REIT gets even closer to being a pure M-M' transaction. Though the land is still out there somewhere, being bought and sold by the REIT, the investor has no experience of it. She invests in farmland exactly as she would invest in any other stock, bond, or derivative. Though financial profits remain rooted in the physical infrastructures and practices, it is in this form that farmland comes closest to being treated as a "pure financial asset."[59]

## From Environmental to Financial Uncertainty

Though financialization has ameliorated some of the risks of investing in farmland and agriculture, it has simultaneously introduced new risks of its own making. An investment conference speaker—a hedge-fund manager whose stocks included operating companies with large amounts of farmland—broached this topic while explaining the chronic underperformance of the South America–focused agribusiness conglomerate Adecoagro. The company's stock price was low, the speaker explained, in part because public markets just don't know how to value companies with large landholdings and in part because of institutional investors' tendency to buy and sell stocks en masse based on a very superficial knowledge of agriculture. With a note of frustration in his voice, he said: "The earnings estimates are still going to be met and beat by the company, but the company is down 35 percent because the money flow is coming out. That didn't happen ten years ago when I started investing in ag. Right now I have to spend time just trying to watch the money flow, because it causes way overvaluation and way undervaluation." At this point, the moderator chimed in: "So it's not just you being smart about your business, it's you being smart about the other people who are being dumb." In other words, investors' herd behavior causes flows of capital in and out of corporate stock based on their changing beliefs. Investor expectations are not rational and are not formed independently.[60]

Publicly traded farmland REITs, because they are straightforward real estate investments, should avoid many of the pitfalls associated with landownership by public companies. Yet these REITs still contend with considerable financial risks. I really understood this for the first time during a 2012 interview with the CEO of one of those early Bulgarian REITs. I was accustomed to talking to more conventional farmland investors and so asked, somewhat naively, about the company's returns. The CEO's answer surprised me. In heavily accented English, he asked: "Do you speak about our profit or about the return of the shareholders?" I suddenly realized that, of course, there were two different types of returns at work—those from farm rents and land appreciation and those from the company's changing share price. Furthermore, there was no guarantee whatsoever that these two types of return would move in tandem with one another. In fact, it turned out in this instance that they had diverged quite considerably, to the detriment of the company's share price. "After the financial crisis," the executive explained, "there is a big difference between the book value of the share and the market value, because the price on the stock exchange is not so closely connected to our profits and activities." In other words, the total value of the existing shares fluctuated with the stock market even though the underlying value of the land and rents had not changed.

The US-based publicly listed farmland REITs have likewise exhibited very disappointing performance. As figure 3.7 shows, by the end of 2018 the stock prices of the two extant US farmland REITs—Gladstone Land and Farmland Partners—had fallen by 20 and 60 percent respectively since their launches. This poor performance is due to a combination of agricultural and financial risks. A slump in agricultural commodity prices has led farmland prices to level off or slightly decline in many parts of the US. Though such ups and downs are typical in land markets, stock-market investors have little inclination or obligation to ride them out. In a 2016 interview with CNBC, Paul Pittman, the CEO of Farmland Partners, emphasized his belief in long-term land-value increase due to fundamental trends: "We think it's a multi-decade trajectory of population growth and GDP per capita growth. If you think there will be no additional demand or declining demand for food, by all means, bail on farmland REITs."[61] The problem is that investors *can* bail on farmland REITs very easily. By making farmland liquid, public REITs give investors the option to sell on the spot. Exhortations to remember the long-term trends may not work well with stockholders, who are notoriously focused on short-term profits.[62]

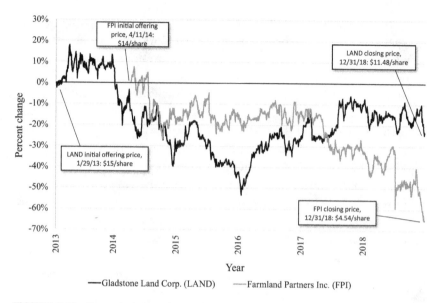

**FIGURE 3.7**    Percent change in stock price of US public farmland REITs since their IPOs, January 2013 through December 2018
*Source:* Image by author based on data from Nasdaq, "Historical Prices."

Financial markets have been particularly unkind to Farmland Partners. In July of 2018, the company's share price plunged 40 percent after an anonymous group of market analysts released a scathing report alleging that the company was inflating its revenues by making loans to tenant farmers who were also members of its own management team and who then recycled the money back to the company in the form of rent. The analysts further alleged that the company had overpaid for some of its properties and was now on the verge of insolvency.[63] Appropriately, the accusers used the moniker "Rota Fortunae," or "wheel of fortune," a Roman symbol of unpredictable luck long associated with gambling and speculation.[64] Rota Fortunae acknowledged that they were shorting Farmland Partners stock—meaning that they had effectively placed bets on the stock price dropping in future.[65] Their report was a perfect example of performativity in action—in declaring publicly that Farmland Partners was in financial trouble, they quickly made it so and no doubt raked in a healthy profit in the process. The company retaliated with a lawsuit accusing Rota Fortunae of conspiring in a "short and distort" attack which was "false, misleading, and defamatory."[66] However, as this book goes to press in 2019, the company's stock has yet to recover much of its lost value.

Not only does the financialization of farmland fail to fully overcome the uncertainties entailed in working with nature, but it simultaneously introduces new elements of financial uncertainty caused by the overreactions, misinterpretations, and prejudices of investors.[67] One of the biggest risks associated with agriculture has long been that millions of farmers globally, all responding to the same price signals, will ramp up production of the same crops at the same time, leading to an oversupply and plummeting crop prices. As agriculture becomes financialized, the herd mentality of investors becomes a concern as well. This problem is not isolated to liquid investments but is endemic to the financial mind-set; the more the owners of land view it as a financial asset, the more likely they are to buy or sell collectively in response to changing interest rates and the performance of alternative investments.[68]

Today's farmland investors may view their properties as a source of financial returns like any other, but they must nonetheless cope with the unruly materiality of agricultural land. Those promoting farmland investment have, in recent years and decades, developed methods for getting around, or even profiting from, the spatial and temporal barriers to capitalist agriculture. These "detours" range from theoretical advances in conceptualizing investment risk to investment vehicles that make possible liquid investments in farmland. These strategies each incrementally advance the financialization of farmland, though they also inadvertently introduce risks of their own making.

In the following chapter, the social complexity of farmland investment is brought into sharper relief through an examination of the political conflicts awakened by large-scale farmland investment in Brazil. The case of Brazil reveals both the possibilities and the limitations of national-level regulation for constraining the financialization of farmland.

# FOREIGN POLITICS

**Quem compra terra não erra (One can't go wrong buying land).**
—Brazilian expression

I had often seen the iconic photographs of Brazilian mega-farms. Rows of gleaming combine harvesters advance across a flat, monochrome landscape that stretches to the horizon, leaving perfect parallel lines in their wake. Sometimes the harvesters move in a V formation, like a flock of bellicose metal geese bearing down on the camera. It wasn't until 2012, however, that I finally had a chance to visit one of these farms as part of an investor tour. The tour group—mostly composed of the representatives of global investment banks and Brazilian financial conglomerates—was shuttled from field to field through a series of carefully orchestrated stops, at each of which a farm employee gave a brief presentation about hectares in production and average yields while gesturing to a glossy informational poster propped on an easel.[1] At one point we stood in a sea of fluffy white cotton, watching enormous harvesters bundle it into bright yellow plastic-wrapped bales that stood out beautifully against the red earth and bright blue sky. The fields were just as interminable as in the pictures, the harvesters just as gleaming. At the end of the day, we trooped inside to watch a sleek slide presentation about the company, including its new initiatives to take advantage of rapidly appreciating land value through targeted land acquisition, transformation, and resale.

Global finance capital is increasingly drawn to Brazilian farmland. It encountered an unexpected hurdle, however, when in 2010 the government placed legal restrictions on the amount of rural land that could be owned by foreigners. Such restrictions are the greatest barrier financial landownership can face. Even more than the moral norms surrounding land or its awkward physical characteristics, laws hold the potential to limit the deepening commodification of land and to ensure that land markets serve social goals. The Brazilian government attempted to do just this—an attempt that, as it turned out, proved singularly unsuccessful. The story of Brazil's attempt to regulate the land rush illustrates the difficulties that arise when governments regulate land acquisitions using traditional tools of territorial control in a context that has been transformed by neoliberal deregulation and financialization. Laws limiting land acquisitions by foreign entities, I argue, while the most readily available tool for stemming the dispossession associated with the global land rush, fall short in the face of supranational capital flows that are relatively untethered to national identity. In Brazil, the attempt to capture highly mobile and fungible global capital within a rigid foreign-domestic dichotomy left openings for some companies to find creative ways around the restrictions. This case suggests that effective laws must address land's growing desirability to investors directly—moving beyond both territorial sovereignty and a fixation on agricultural productivity.

## A Primer on Land Relations in Brazil

Before delving into the heated national politics surrounding Brazil's regulation of foreign farmland acquisitions, however, we must first know something about the meaning and history of land in Brazil. Brazilian landownership is extremely concentrated. The breathtaking scale of operation I witnessed on the farm tour is made possible by one of the most inegalitarian landownership structures in the world. The Gini coefficient (a measure of inequality in which 0 signifies the least inequality and 1 the most) for Brazilian land distribution is a staggering .85. Farms smaller than 10 ha make up 47 percent of the total number of farms but occupy only 2.7 percent of the total area of rural establishments, while farms over 1,000 ha make up less than 1 percent of the total number of farms but occupy 43 percent of the area.[2] This unequal distribution has major consequences. First, it means that Brazil's development into an agricultural superpower since the 1970s has largely benefited elite individuals and enormous operating companies like the one I visited, rather than bestowing widespread benefits on small and medium-size producers. Second, it makes land an intensely political issue in Brazil—the center of a heated dispute over agrarian reform. Nowhere does the social life of land involve more drama than in Brazil.

The roots of this land concentration stretch back to the colonial period. From the sixteenth century to the eighteenth, the Portuguese monarchy controlled its massive colony through a semifeudal system of land grants known as captaincies. The heads of these captaincies were tasked with colonizing Brazil and, among other powers, had the right to grant tracts of land, called *sesmarias*, to individuals. Land grants under the *sesmaria* system conferred noninheritable use rights, rather than private ownership, as all land was considered property of the crown. The *sesmarias* were at least nominally contingent on productive use of the land, so they were granted only to people of means, generally friends of the crown. The result was that a limited number of elite families came to control most rural land, in the form of enormous coffee and sugar plantations (subsequently termed *latifúndios*) or cattle ranches. Meanwhile, the majority of the population remained landless agricultural laborers or toiled on plots too small to suffice. This landownership structure ensured a labor supply for the *latifúndios* by separating workers from the means of production.[3]

**FIGURE 4.1**   Brazil's regions and states

*Source:* Bill Nelson. Adaptation of "Brazil Labelled Map" by Felipe Menegaz. Used under CC BY-SA 3.0.

**FIGURE 4.2**  Brazil's biomes
*Source:* Bill Nelson. Adaptation of Brazil Travel, "Map of Biomes of Brazil." Used under CC BY-SA 4.0.

Brazil gained its independence in 1822, bringing an end to the *sesmaria* system but not to the unequal distribution of land. The void left by the end of the *sesmaria* system was filled, in 1850, by a new law that strengthened the elite's grip on rural land.[4] The Land Law of 1850 had several important aspects. First, it formalized the potential for rural land to become private property, thereby creating an official land market for the first time. Existing land claims, including the former *sesmarias*, could now be legally registered as the private property of their current owners.[5] Second, the Land Law stated that public lands could be acquired only through purchase and not through squatting, thereby putting landownership out of reach of the vast majority of Brazilian peasants. At a time when a growing abolition movement threatened to remove slavery as a source of plantation labor, this law ensured a steady supply of wage workers in the form of poor European migrants and other Brazilian peasants who now had little chance of becoming landowners in their own right.[6] The significance of this law is underscored by the very different course of action taken by another agricultural giant: while the Brazilian government was busy cementing its colonial legacy of unequal land distribution, the US government was shoring up the place of family farmers by giving them free land under the 1862 Homestead Act.

Crucially, while the Land Law of 1850 made it hard for poor people to acquire land through squatting, the wealthy still had means of accumulating public lands. The Land Law was ineffectually and arbitrarily enforced, so those with means could still add to their landholdings through a practice known as *grilagem*—land grabbing cloaked in document fraud.[7] *Grilagem* involves creating a "legal" land claim through forged titles, phony paper trails, and, often, hired gunmen. This process was aided by Brazil's unusual land registration system, which operates through public registry offices known as *cartórios*. The *cartórios* were, until recently, hereditary, and they are notoriously inefficient or even corrupt—all of which facilitates land fraud.

Over the tumultuous decades that followed, little occurred to correct the deeply unequal distribution of land in Brazil. Projects to modernize the countryside during the early twentieth century led to the construction of highways and railroads, which brought speculative profits to the elites who already owned land but did little to broaden the ownership base.[8] Toward the middle of the twentieth century, however, peasant resistance began to build, and agrarian reform became a point of overt political tension. Rural unrest increased throughout the 1940s, 1950s, and 1960s, as did rural organizing through newly created peasant leagues, which demanded land redistribution, and rural unions, which demanded better contracts for rural workers.[9]

In 1961, with the election of left-wing president João Goulart, it began to look like the agrarian reform agenda had a chance of success. Goulart showed support for rural activists by introducing progressive legislation that facilitated rural unionization and by drafting a Land Statute that included land redistribution. However, these moves were seen as a threat by both Brazilian elites—whose social power is deeply tied to their hold on the nation's land base—and the urban middle class, who, in the Cold War context, feared any shift in the direction of communism. Ultimately Goulart's support for agrarian reform contributed to his 1964 overthrow by an elite-backed military coup. Control over land lay at the heart of the Brazilian power structure, and elites would not give it up without a fight.[10]

In the two decades of military dictatorship that followed Goulart's ouster, land reform again became a remote dream. Once in power, the military cracked down on rural mobilization, arresting and torturing peasant league leaders.[11]

The military regime produced a watered-down Land Statute, which jettisoned the emphasis on land reform and attempted instead to simultaneously diffuse peasant unrest and secure control over Brazil's vast territory through a program of internal colonization of the frontier.[12] It embarked on a program of infrastructure development, building roads and ports in the scantily populated North and Northeast regions. It also sought to expand the reach of capitalist agriculture into the countryside. In an effort to bring the fruits of the "Green Revolution" to Brazil, it founded a national agronomic research institute, the Brazilian Agricultural Research Corporation (EMBRAPA), which used modern agricultural technology to expand the agronomic potential of the infertile *cerrado*—the vast tropical savannah that dominates the center of Brazil.[13] This agricultural development campaign was highly successful; Brazil, long a producer of traditional tropical export crops like coffee, sugar, and rubber, suddenly emerged as a major contender in global markets for staples like soy and wheat. Overall, the military regime's approach to rural Brazil was one of "conservative modernization"— rather than altering rural property relations, it encouraged the adoption of input- and capital-intensive agricultural production and the conversion of peasants into wage workers.[14]

Yet running in parallel to this long history of enabling elite and capitalist land concentration, Brazil also has a strong legal and cultural tradition of asserting that land must serve a *social function*. The doctrine of the social function of land contends that the state's protection of private property is justifiable only when the land is used in such a way as to benefit society as a whole. Many countries, particularly in Latin America, have a tradition of treating property as a means to achieve policy goals rather than as an unassailable individual right.[15] Although the social function doctrine is most commonly associated with the nineteenth-century French scholar Léon Duguit, some version of it has been present in Brazilian land law for much longer. Under Portuguese rule, *sesmarias* could be revoked if not put to productive use within six years, though this aspect of the law was rarely enforced.[16] The Land Law of 1850 also gestured toward the social function of land, stipulating the legitimation only of those occupations distinguished by effective cultivation and habitual residence. An explicit statement on the social function of land was present in the Constitutions of 1934 and 1946 (though it was absent from the Constitution of 1937).[17] Even the 1964 Land Statute allowed for the compensated expropriation of unproductive *latifúndios* in the "social interest," though little redistribution occurred in practice.[18]

When democracy returned to Brazil in the mid-1980s, peasant movements rapidly reemerged and renewed their demands for agrarian reform, seizing upon the social function of land as their primary weapon. The military dictatorship gave way to a democratically elected government in 1985, and a new Brazilian Constitution was passed in 1988. This constitution reaffirms that land must fulfill a social purpose—defined by environmental preservation, worker well-being, and (above all) productive use—or risk government expropriation. Thanks to vigorous lobbying by large landholders, however, the constitution also protects private property by guaranteeing that the government must compensate the owner whenever it expropriates land.[19]

Wielding this clause of the new constitution, the movement for agrarian reform came out from the shadows. It was led by the Landless Workers' Movement (Movimento dos Trabalhadores Rurais Sem Terra, MST), a vibrant new force representing the rural poor, which began organizing the occupation of unproductive *latifúndios*. During the 1990s the MST made significant headway, eventually convincing the administration of Fernando Henrique Cardoso (in office from 1995 to 2003) to enact a significant program of land reform for the first time, despite the neoliberal tenor of this government. The MST achieved heightened social legitimacy under the left-wing presidency of Luiz Inácio Lula da Silva, known as "Lula" (2003–2011), a former union leader and member of the Workers' Party (Partido dos Trabalhadores, PT) who extolled the importance of agrarian reform. Under the Lula administration and that of his successor Dilma Rousseff (2011–2016), the MST experienced unprecedented verbal and financial support, though actual progress in redistributing land was slow. Overall, the 1990s and early 2000s were the heyday of land reform in Brazil, a time when the tide of land concentration finally seemed to be turning.

Today, land policy in Brazil remains a highly contested political terrain. On the one side, the MST, the National Confederation of Agricultural Workers (Confederação Nacional dos Trabalhadores na Agricultura, CONTAG), and other peasant-based groups continue to demand a redistribution of farmland. On the other side, elite interest groups like the Rural Democratic Union (União Democrática Ruralista, UDR), the Agriculture and Livestock Confederation (Confederação da Agricultura e Pecuária do Brasil, CNA), and their representatives in Congress, the Rural Caucus (*bancada ruralista*), exert their considerable influence to preserve the status quo. Since the mid-2000s, the land reform movement has lost some of its initial steam; membership in land reform organizations has declined, as has public support for the cause.[20] The problem, however, has not diminished: rural Brazil is still characterized by extreme concentration of landownership. It is perhaps inevitable, then, that although agrarian reform is no longer a centerpiece of the national political agenda, it simmers constantly below the surface. Government regulation of land, meanwhile, remains contradictory and ineffective. Inefficiency in the *cartório* system has led to many overlapping land registrations, the government has been unsuccessful at enforcing payment of rural land taxes, and the national cadastre is not sufficiently modernized, with the result that the location and extent of both private properties and public land is uncertain.[21] All of this has the result of enabling continued *grilagem*.

Brazilian elites view land not only as a source of political power, but as a way to store and increase their wealth, and this use has greatly influenced historical land price movements.[22] During the 1970s, the military government's modernization drive created considerable opportunities for real estate speculation by elites, often via *grilagem*.[23] Both because of this speculation and because of rapid gains in productivity, land prices rose rapidly during this period (see figure 4.3). They reached a new plateau in the early 1980s but then, owing to macroeconomic insecurity, entered a period of dramatic price fluctuations that lasted from the mid-1980s to the mid-1990s. During this period, chronic inflation reduced confidence in the economy, while the rocky return of democracy brought political instability. Every time an attempt to stabilize the economy failed, land became more valuable as a haven for capital, and prices shot up.[24] After the currency was finally stabilized in the mid-1990s, the price of land—no longer inflated by economic anxiety—dropped dramatically, demonstrating the extent to which land had been acting as a store of wealth during the troubled decades before.[25] In the early 2000s, however, the price of land began to edge up once again.

**FIGURE 4.3**  Average price of Brazilian cropland, inflation-adjusted, 1966–2014[26]

Source: Image by author based on data from Fundação Getulio Vargas Instituto Brasileiro de Economia, "FGV Dados" (years 1977–2014) and Plata, "Mercados de terras no Brasil" (years 1966–1976).

## Foreigners in Our Midst: Brazil's Regulatory Response to the Land Rush

Brazil was among the first countries to feel the effects of the global land rush. As crop prices climbed throughout the mid-2000s, Brazilian farms became a hot commodity. With the growing number of large-scale land acquisitions in Brazil, critics adopted the term *estrangeirização* or "foreignization" to refer to the process. This designation conveniently differentiates the land rush from Brazil's home-grown variety of land grabbing, *grilagem*, but it also glosses over a lot of complexity in the types of actors drawn to Brazilian land.[27] This emphasis on foreign nationality does seem relevant when describing the pattern of government-backed enterprises seeking land abroad to ensure food security at home—what Philip McMichael terms "agro-security mercantilism"[28]—but this is not the predominant pattern in the Brazilian land rush. The primary exception to this rule was Chongqing Grain Group's plan to buy 100,000 ha of *cerrado* farmland, funded largely by the China Development Bank and the China National Agricultural Development Group Corporation, a deal that received a lot of media attention and was ultimately canceled.[29]

Instead, the Brazilian land rush is primarily motivated by profit, and few of the leading actors are easily identified by any single national origin. International finance capital figures prominently among the protagonists. For instance, Vision Brazil Investments, a São Paulo–based asset management company created in 2006 by former Bank of America executives to serve foreign institutional investors, sponsored a fund called TIBA Agro, which quickly amassed over 300,000 ha of Brazilian farmland.[30] Brookfield Asset Management, a global investment company with over $200 billion in assets under management, sponsored two funds—Brookfield Brazil AgriLand Fund I, which closed in 2011 after raising $330 million in capital, and AgriLand Fund II, which closed in 2015 at roughly $500 million.[31] Nuveen, a subsidiary of pension-focused asset manager TIAA, held 738,000 ha of Brazilian farmland as of 2018, including properties purchased via the two global farmland funds it manages on behalf of third-party investors (see introduction).[32]

Investment also frequently occurs via Brazil's large landowning agribusinesses, many of which—including the one I visited—are publicly traded. One example is Adecoagro, the publicly traded agribusiness whose low stock market valuation was mentioned in the previous chapter. Adecoagro (registered in Luxembourg) has a holding company that as of 2018 owns 226,000 ha across South America. Its major shareholders include Qatar-based Al Gharrafa Investment Company, the Dutch pension fund service provider Stichting Pensioenfonds Zorg en Welzijn, and US-based investment management companies EMS

Capital and Route One Investment, as well as many other institutional investors.[33] Another publicly listed agribusiness is BrasilAgro, which bills itself as both agricultural operator and farmland development company. Just over 40 percent of BrasilAgro's shares are held by the Argentine agribusiness Cresud, while the remainder are in free float (meaning that they are openly traded on the stock market). As of 2017, BrasilAgro owned 200,000 ha across Brazil and Paraguay.[34] Being traded on the stock market makes companies like Adecoagro and BrasilAgro an easy way for investment capital to access land markets, but it also means that the precise breakdown of their ownership varies from day to day.

To describe these finance-backed funds and companies as "foreign" is not entirely *inaccurate*, but it is certainly *insufficient*. Rather than being transferred from domestic entities to unitary foreign entities, Brazilian land frequently comes under the control of capital whose national affiliation is either unstable, multiple, nontransparent, or simply designed to ensure preferential tax treatment. Yet while the threat posed by the land rush in Brazil was fundamentally financialized and deterritorialized, the government regulatory response focused on *foreigners* (rather than investors writ large) and on land as *territory* (rather than commodity).[35]

In Brazil, restrictions on foreign landownership have gone through several sets of modifications that have tracked the country's changing political-economic climate. These restrictions were created under the military dictatorship (1964–1985), largely repealed under the Cardoso administration's neoliberal push (1995–2002), and then reinstated under the PT administrations of Lula and Rousseff (2003–2016) (see figure 4.4 for a full timeline of key legal events). These fluctuations can be seen as the outcome of a constant Polanyian tension between efforts to free land markets and efforts to make them serve a social function. Regulating foreigners simply serves as an easy—though, I will argue, fairly ineffective—way to achieve broader social goals when it comes to land use.

Brazil's regulations on foreign land acquisitions go back to the 1960s. The military government saw foreign investment as a cornerstone of its development strategy[36] and was therefore initially highly permissive of rural land purchases. However, after a spate of critical media articles on foreign land acquisitions were published in 1967, the National Congress created a commission to inquire into the phenomenon.[37] The commission's findings, known as the Velloso Report, revealed that roughly 20 million hectares of Brazilian farmland had been acquired by foreigners, particularly US companies and individuals, much of it within the Amazon region.[38] In response, the military government passed Law 5.709 in 1971, which limits the amount of land that foreigners can own—both as a group and by nationality—and requires approval from the national government for foreign land acquisitions over a certain size (see table 4.1 for an updated

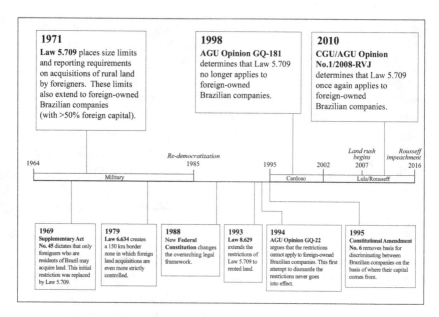

**FIGURE 4.4**    Timeline: Brazilian regulation of foreign acquisitions of rural land
*Source:* Image by author.

version of these restrictions).[39] Crucially, the law specifies that *these restrictions and requirements apply not only to foreign individuals and companies but also to Brazilian companies with majority foreign ownership.*[40] Despite Brazil's general openness to foreign investment, these restrictions fit the nationalistic ethos of the military regime, which was then deeply engaged in efforts to consolidate control over Brazil's sprawling and sparsely populated national territory.[41] However, as the regime changed, so did the government's attitude toward foreign investment.

The legal framework restricting foreign investment in land was largely dismantled during the extensive neoliberal reforms of the 1990s. In 1995, a decade after democracy returned to Brazil, Fernando Henrique Cardoso was elected president, largely on the strength of his success at stabilizing inflation in his previous post as finance minister. Having already achieved this cornerstone of neoliberal reform, President Cardoso set about further implementing the so-called Washington Consensus. His administration privatized public utilities and transportation infrastructure, cut government expenditures, and liberalized everything from domestic finance to foreign trade.[42] Eventually, and perhaps inevitably, the restrictions on foreign land acquisitions also came under fire. In 1998, the attorney general issued an opinion reinterpreting a key provision of Law 5.709. It stated that, in light of changes to the constitution, *the law now applied only to foreign individuals and companies, and* not *to foreign-owned Brazilian companies.*[43] Though Law 5.709 remained on the books, this legal interpretation rendered it inoperable and essentially obsolete. A foreign company could now acquire unlimited Brazilian land without any government approval simply by opening a Brazilian subsidiary.

As the "global land grab" began to make headlines internationally, however, in 2007 and 2008, this laissez-faire policy toward foreign land acquisitions was called into question. One of the immediate catalysts of the reregulation was an incident involving Stora Enso, a multinational cellulose and paper company headquartered in Finland. Stora Enso was found to have been using Brazilian front companies to purchase large amounts of land near Brazil's southern border.[44] The MST brought this acquisition to national attention in March 2008, in language highlighting the threat to national sovereignty. "Attention nationalistic military: help us expel the Finnish from the border," quipped MST leader João Pedro Stedile.[45] A few days later, as hundreds of MST women occupied one of Stora Enso's properties, the president of the governmental body responsible for regulating foreign investment in rural land—the National Institute for Colonization and Agrarian Reform (Instituto Nacional de Colonização e Reforma Agrária, INCRA)—appeared before a hearing of the Brazilian Senate Agriculture Commission and spoke about the lack of regulatory control over foreign land acquisitions. Though he specifically mentioned the threat of pension fund investments in land, he framed the regulatory challenge primarily in terms of national sovereignty: "The central point for discussion on this topic for us is national sovereignty. It's not xenophobia. It is the country knowing . . . what is the destiny of the territory."[46] This framing of the problem is curious because Stora Enso's nationality is by no means easy to pin down. It has major shareholders among Swedish and Finnish institutional investors, including a Finnish state-owned investment company but also several private pension funds, insurance companies, and foundations. Additionally, because it is a publicly traded company with over 70 percent of its stock in free float, the geographic origins of its capital are always in flux—some of it presumably originating within Brazil itself.[47] Stora Enso's acquisition was a good example of land *financialization*, yet it was primarily treated—discursively and legally—as a case of *foreignization*.

In the months that followed, concern about foreign land acquisitions increased. Brazilian newspaper headlines reported that foreigners were buying the equivalent of "six Monacos" per day in the Brazilian countryside or, later, "22 soccer fields" per hour.[48] INCRA commissioned several studies of the situation, though these mostly served to underscore how little the government knew about the extent of foreign landownership in Brazil;[49] since their data only included acquisitions that were officially registered as foreign, they had likely captured only a fraction of the land acquisitions from 1998 to 2010, and the accuracy of reporting before that period was also very open to doubt. Finally, in 2010, just as Lula's term was coming to an end, the attorney general's office issued a new interpretation of the law. This interpretation reverted to the pre-1990s legal framework, meaning that *Brazilian companies with majority foreign capital were once again treated as foreigners under Law 5.709.*[50] The provisions of Law 5.709 are summarized in table 4.1. They are based on an administrative unit called the Indefinite Exploitation Module (MEI), the size of which varies across the country from as low as 5 ha to as high as 50 ha, depending on soil quality, transportation infrastructure, and other factors.

How this saga of de- and re-regulation will end is as yet unknown. In 2016, President Rousseff was impeached in a politically charged corruption case, which greatly altered the political landscape. For many on the political left, this impeachment was simply a coup by another name. Conservative politicians who had failed to gain power through democratic means—and many of whom were themselves caught up in corruption scandals—found it more expedient to remove Rousseff through impeachment than to wait and attempt to gain executive power at the ballot box. Much like the coup against Goulart, the coup-by-impeachment of Rousseff replaced a democratically elected, left-leaning administration supported primarily by the working class with a right-wing administration backed by the upper middle class and elites. Also like the coup against Goulart, Rousseff's impeachment was followed by a rapid abandonment of the agrarian reform agenda. While the Rousseff administration frequently disappointed proponents of agrarian reform, the administration of President Michel Temer, who replaced her, was explicitly allied with the conservative *bancada ruralista*. Under the Temer administration, resettlements of landless families ground to a virtual halt, indigenous land claims lost recognition, and environmental protections were weakened.[51] The restrictions on foreign land acquisitions were also immediately placed in jeopardy once again. Though the situation has yet to be resolved as this book goes to press in 2019, *ruralistas* within the Brazilian Congress are again pushing for deregulation of foreign land acquisitions.[52] If their efforts succeed, the MST has already threatened occupations of foreign-owned properties.[53]

**TABLE 4.1**  Restrictions under Law 5.709

| |
| --- |
| Any project with over three MEI, in continuous or discontinuous plots, requires prior approval from INCRA. Acquisitions with over 50 MEI (in the case of individuals) and 100 MEI (in the case of companies) require approval from the National Congress. |
| Total land owned by all foreigners cannot exceed 25% of a single municipality. |
| Land owned by foreigners of the same nationality cannot exceed 10% of a single municipality. |
| *Cartórios* must keep a special register of foreign acquisitions. They must report on these to INCRA every trimester. |

*Source:* Created by author.

The history of Brazil's attempts to regulate foreign land acquisitions is one of pendulum swings between more and less regulated land markets. There is, as Polanyi described, a perpetual tension between efforts to disembed land markets from all regulatory control and efforts to re-embed them in society via protective measures. What is striking in this case, however, is how dependent the regulatory framework is on the idea of the foreigner. The deregulation of the 1990s allowed for the unfettered capital flows required by neoliberal globalization but never ceased paying homage to a conceptual division between the foreign and the domestic. Rather than abandoning the foreign/domestic dichotomy, the deregulation was achieved simply by moving the border between the two categories, making it exceptionally easy to be Brazilian in the eyes of the law. Likewise, when the government wanted to take action against the land rush in 2010, it just moved that border back again, reclassifying foreign-owned Brazilian companies with the foreigners. The legal response to the land rush could have hinged on the issue of corporate landownership or on the increasing agricultural consolidation likely to result from large capital flows into rural land markets, but instead it was constrained by an exclusive preoccupation with *foreigners.*

## The Foreigner as Proxy in the Politics of Productivity

Why was Brazil's political response to the land rush limited to a policy of restricting land purchases by foreigners? I will argue that the foreigner made for a useful proxy—a way to regulate the ongoing concentration of landownership in the hands of agribusiness at a time when the Brazilian government's tool kit for doing so is very limited.

Carlos, a top executive at a Brazilian operating company with major landholdings, complained that the restrictions were only masquerading as being about foreigners, when in fact they were part of an ideological struggle over the structure of Brazilian agriculture:

> [The government says that] the land used to be used by smallholders, and if you have those big guys, they're going to concentrate land and buy more land and become big producers, which is bad for the country. So the discussion here, it's very easily not [about] foreign capital. It's capital and labor, actually. That, for me, is the discussion that is being proposed in Brazil today, is that of capital and labor in agriculture. Big farmers, small farmers; that is the discussion we are having right now and it's masked as if it were foreign capital.

He went on to say that if you took the attorney general's opinion, which rein-stated the restrictions in 2010, and replaced the term "foreign capital" with the word "capital" throughout the document, the entire thing would still make sense, because the arguments are really about preventing the concentration of land-ownership and reserving land for smallholders. He was exaggerating, but he had a point. The attorney general's opinion listed nine reasons for restricting foreign land acquisitions, of which only the last one—"acquisition of land in the frontier zone putting national security at risk"—related specifically to foreigners.[54] The remaining eight could have applied equally to any influx of private capital into land markets; they related to illegal activity, environmental protections, and the social relations of landownership—in short, they suggested the need to re-embed farmland markets by making them serve a social function.

Almost everyone I interviewed believed (for better or for worse) that a pri-mary motivation of the reregulation was a desire to limit land price increases in order to facilitate agrarian reform. Indeed, the attorney general's opinion gave this as the second reason for the restrictions, warning of "irrational land price appreciation and incidents of real estate speculation generating an increase in the cost of the expropriation process for agrarian reform, as well as a reduction in the stock of land available for this end." Since the government must pay market-rate compensation whenever it expropriates land, officials within the Lula and Rousseff administrations were worried that increasing demand for land would drive up land prices, reducing the pace of land redistribution. It is probably not a coincidence that INCRA, the government entity that regulates foreign land acquisitions and pushed for the reregulation, is the same entity charged with administering the national program of land reform.

A desire to foster agrarian reform was explicit in my interviews with pro-restriction government officials. To give just one example, I asked an official at the Ministry of Agrarian Development to explain the reasons for the restrictions, and he first listed the issue of national sovereignty, but this was immediately fol-lowed by, "The second question was the fear that these massive investments of money into agribusiness land is contributing to a hike in the price of land." He went on to say,

> It is absolutely evident that massive investment into agricultural land in Brazil has two main effects. First, the price of land. Second, which is not in terms of importance, is concentration of land. I mean, these guys don't come here to buy 20, 25 hectares. Or even 100, or even 1,000. They come here to buy something like 20, 25, 30, 35 thousand [hect-ares], okay? And of course, they buy the best land. . . . And so this is a serious concern. We have a program, which is the program of land

reform, which is at a low in our country. And we are the ministry that is [charged] with the implementation of this program. And so we have to do it. In these circumstances—economic circumstances, investment circumstances—it is much more difficult.

Though agrarian social movements were disappointed with the PT government's track records of expropriating and redistributing land, the administration's stance was still officially in favor of land reform and family farming; restricting foreign investment in land was an opportunity to be seen making some headway in these battles.

If a primary goal was to limit an influx of corporate capital into land markets and promote a more equitable distribution of land, why not just say so? Why wrap the issue up in talk of foreignness and national security? The answer to this probably lies largely in the very narrow ways in which the "social function of land" has come to be legally defined in Brazil. During the drafting of the 1988 constitution, the definition of *social function* was intensely debated, since it would eventually serve as the basis for determining which land could be expropriated for land reform. Quite inclusive definitions were considered, but ultimately the very restricted definition advocated by the rural elite prevailed. As a result, the social function of land as defined in the constitution includes the treatment of the environment and labor, but the one element that trumps all others is productive use.[55] In the words of Sérgio Sauer and Sérgio Leite, this narrow definition of social function restricts "land to an economic dimension," minimizing its importance "as a means and place of production as well as a source of identity, self-recognition, historical occupation, and life."[56] The myriad social and cultural meanings of land are reduced to a single, quantifiable dimension. If social function is purely a question of economic productivity, then Brazilian land is fulfilling its social function (and therefore cannot be expropriated) even if it is entirely owned and operated by large-scale agribusinesses backed by pension fund capital from abroad.

Whereas the neo-Lockean idea that land use and transformation justifies ownership is prevalent throughout the West, in Brazil it is enshrined in law as virtually the only grounds for agrarian reform. This leaves little leeway for policy makers to address farmland financialization. As discussed in chapter 2, major financial investors rarely leave their properties idle. They speculate on rising land prices, but this speculation should not be conflated with disuse; if anything, their overriding focus on land price appreciation stimulates them to use land intensively and put it to higher-value uses whenever possible. Reduced to productive use, the social function of land is stripped of much of its political power. Thus defanged, it can be used comfortably even by those who favor ongoing land accumulation by the owners of capital. For instance, deputy Marcos Montes[57]—a member of the *bancada ruralista* and an opponent of the 2010 restrictions—told me that "foreign investment is welcome in the country as long as it isn't speculative." When I followed up by asking what speculative investment means in this situation, he explained: "Speculative investment in this case is a venture where the [social] function of land is not adequate. . . . [You] don't do with the land what we want, which is produce food, produce energy, and you leave the land standing there for the purposes of appreciation." Defined so narrowly, the social function of land is neutralized as a political tool, leaving opponents of ongoing landownership consolidation with little recourse but to turn to proxies such as foreignness. Hamstrung by this restrictive politics of productivity, the government's only recourse was its age-old right to protect its territorial sovereignty.

Addressing the land rush as an issue primarily of a foreign threat to territorial sovereignty was expedient, particularly in Brazil, which has long harbored fears of losing control over its vast territory—either to neighboring countries that might nibble away at the borders or to Northern environmental organizations attempting to "internationalize" the Amazon for conservation purposes.[58] It was a framing of the problem that groups from both sides of the political spectrum could potentially support, with even some conservative groups such as the CNA initially expressing mild approval.[59] After the reregulation had taken place, even those working to repeal the new law worked within this framing of the problem; there was a foreign threat, opponents argued, but it had been defined too expansively. During 2011 and 2012, the debate over foreign land acquisitions played out in a temporary subcommittee of the Agricultural Committee of the Chamber of Deputies (the lower house of Brazil's Congress). I attended some of the committee meetings, which often took place in a cramped back conference room belonging to the Agriculture Committee where the deputies squeezed around a table in the middle of the room with aides standing over their shoulders and reporters crowded around the walls. As the end of its term neared in 2012, the subcommittee had to choose between two alternate proposals for legislation. The first, sponsored by Deputy Beto Faro, the only PT member on the subcommittee, would have essentially kept the restrictions in place at their current levels—with foreign companies defined by majority foreign capital. The second proposal, and the one that ultimately prevailed within the subcommittee, was presented by Deputy Marcos Montes, one of the many *ruralista* subcommittee members. It framed the foreign threat more narrowly, proposing to once again remove restrictions on land acquisitions by foreign-owned Brazilian companies while leaving restrictions on sovereign wealth funds, state-owned companies, and foreign NGOs.[60]

Opponents of the reregulation burnished their nationalist credentials by pointing to the "genuine" threat posed by Chinese investors. In Brazil, as in many parts of the world touched by the global land rush, Chinese-backed land purchases received a vastly disproportionate amount of attention from the media, activists, and politicians.[61] Unlike white investors, the Chinese are depicted as the racialized "other";[62] they are often perceived as acting en masse and in culturally prescribed ways, an assumption that is never extended to the Americans or Europeans acting in Brazilian land markets. Deputy Homero Pereira, the president of the special subcommittee on foreign land acquisitions—and a staunch *ruralista*—told me in an interview that his primary concern was with Chinese and Indian land acquisitions. "I have information that the Chinese invaded Madagascar and practically already bought the entire country," he said, in an apparent reference to South Korean company Daewoo's canceled land acquisition in Madagascar (see introduction). Everyone, it seemed, could agree that the Chinese were foreigners; the only question up for debate was whether Brazilian companies owned by American or European pension funds should also be considered foreign.[63] Constrained by the politics of productivity, the complex issue of financialized land acquisitions by transnational companies was reduced to a simple question of who really counted as "foreign."

## The *Jeitinho Estrangeiro*

Limited to a conceptual and legal tool kit unchanged since the 1970s, the reincarnated law is an incomplete measure for addressing the transnational and financialized threat posed by the land rush. The reregulation had an immediate but uneven impact. Chongqing Grain Group, whose undisputed affiliation with a single foreign nation made it one of the few entities adequately described by the legislation, canceled its plans to acquire farmland and announced that it would instead build a soy-crushing plant in Brazil.[64] Among the more plentiful private investors, however, the regulation's success was more equivocal. In the immediate aftermath of the reregulation, several planned Brazilian farmland funds were postponed or raised less than their target, and agribusiness associations released reports on the amount of foreign investment being forfeited as a result of the regulatory change.[65] The reregulation caused particular uncertainty for publicly traded operating companies, given that the extent of their foreign capital changes from day to day as stocks are bought and sold. Opponents of the restrictions also argued that they created an obstacle to farm financing by making foreign banks and trading companies reluctant to accept land as collateral on agricultural loans.

However, interviewing fund managers and company executives in São Paulo in 2011—a year after the reregulation—I found that the new regulations were by no means an impenetrable barrier to farmland acquisitions. Few interview participants were willing to admit to me outright that their company was still buying farmland with foreign capital, but several were comfortable mentioning other companies that were doing so or telling me about how they might hypothetically do it in a level of detail that betrayed serious consideration. Methods for getting around the new restrictions were also discussed openly at several of the agricultural investment conferences I attended. At a conference in 2013, an audience member asked a Brazilian farmland fund manager to explain the effects of the new restrictions. He replied, "I can talk all day naming and explaining the details of the ways that foreign investors can buy and operate land in Brazil." His comment evokes the cliché of the *jeitinho brasileiro*—the "Brazilian way"—a much-vaunted national ability to creatively overcome problems, often by bending the rules. In this case, the "ways" frequently exploit the discrepancy between the flexibility of global capital and the rigidity of the foreign/domestic dichotomy on which the law is based.

The largest, most publicly visible companies were probably least likely to be able to get around the restrictions, yet most found creative ways to comply without substantively altering their investment plans. One option was for a company to migrate across the boundary line separating the foreign from the domestic by shifting holdings between parent companies and subsidiaries. An example of this strategy comes from Agrifirma, a farmland-focused agricultural operating company created by British financiers in 2008. In September 2011, a year after the foreign-ownership regulations went into effect, the company underwent a major restructuring.[66] It received an infusion of capital from BRZ Investimentos, a Brazilian asset management company. However, this new Brazilian capital alone would not have been sufficient to get Agrifirma over the hurdle (50 percent domestic capital) required to continue buying large tracts of land without government approval. In order to complete the process, the company changed its name to Genagro and created a subsidiary company called Agrifirma Brasil Agropecuária (ABA), which alone received the new shot of domestic capital. The company carefully distributed its existing assets between the parent company (Genagro) and the subsidiary (ABA) to ensure that BRZ Investimentos would hold the majority share in ABA. Roughly half of Agrifirma's original assets, including one fully developed farm, cash, and financial assets, remained with Genagro, while three relatively undeveloped farms were transferred to ABA. This asset reshuffling allowed ABA—now a "Brazilian-controlled" company—to continue pursuing farmland acquisition, development, and resale within Brazil unimpeded. Leaving the undeveloped farm properties under ABA made sense, because they had the greatest potential to increase in value over the coming years and because they were worth the least on paper and therefore didn't outweigh the new Brazilian capital infusion on the company balance sheets.[67] It is hard to imagine that the division between Genagro and ABA is anything more than cosmetic, but it allowed the company to escape the restrictions.

Other large companies adapted to the law in different ways. Before the new restrictions went into effect, the publicly traded Brazilian agribusiness SLC Agrícola was shopping around a new farmland acquisition and development company called SLC LandCo to foreign investors. The new regulation at first put these plans on hold, but in 2012 SLC announced that LandCo would receive a $240 million investment from the British asset management firm Valiance. After a careful distribution of existing farm assets, LandCo became just over half Brazilian-owned—by the time the investment was complete, SLC Agrícola would hold 50.6 percent of the shares and Valiance would hold the remaining 49.4 percent. Interestingly, however, the parent company, SLC Agrícola, is also just over half Brazilian-owned—51 percent of its shares are owned by one of its Brazilian founding families, while the other 49 percent are free-floating.[68] This meant that the Brazilian half of LandCo's ownership was also only half Brazilian. One therefore might assume, given the laws governing the multiplication of fractions, that LandCo could only guarantee that slightly over a *quarter* of its capital came from Brazilians. However, an SLC Agrícola executive I interviewed assured me that this ownership structure is perfectly adequate to comply with the law.

Radar Propriedades Agrícolas, the joint venture between US pension fund TIAA and Brazilian sugar-alcohol agribusiness Cosan, approached the new regulatory framework by disputing the legal meaning of a foreign-controlled Brazilian company. TIAA (then TIAA-CREF) supplied 81 percent of the capital for Radar, but Cosan has more voting positions on the board. In a 2011 interview, one of the company's legal representatives explained that—in the company's opinion—the restrictions should actually be interpreted in light of Law 6.404, passed in 1976, which defines shareholder control in terms of the number of voting seats on a company's board rather than in terms of majority capital as stated in the 1971 law. To be on the safe side, however, TIAA and Cosan also formed a separate company in 2012—Tellus Brasil Participações Ltda—to channel foreign capital into Brazilian farms. On paper, Tellus is majority Brazilian-owned: 51 percent is owned by Cosan and 49 percent by subsidiaries of the global farmland investment fund TIAA-CREF Global Agriculture. However, the capital that Tellus uses to buy farm properties is largely supplied by subsidiaries of TIAA and the farmland funds it manages via debentures, a type of unsecured debt instrument that can often be converted directly into company stock.[69] Now in compliance with the letter of the law, though certainly not its spirit, TIAA's Brazilian farmland holdings climbed from 104,359 ha in 2012 to 738,907 ha in 2018.[70]

For smaller, less heavily scrutinized companies, the options for getting around the restrictions are more varied. One approach is to disguise foreign capital behind a Brazilian front man (known in Brazil as a *laranja* or "orange"). The use of front men is a strategy as old as the hills—or in this case the very flat *cerrado*—a longtime tool of land-grabbing Brazilian elites. It shows that the state is not the only actor repurposing existing institutions to deal with increasingly transnational capital flows; those attempting to evade state regulation are also drawing from the existing repertoire of *grilagem*. In this approach, the land is sold to a foreigner or foreign-controlled company using private legal contracts, while it remains officially registered to a Brazilian—either the former owner or a third party. A more sophisticated variant of this approach, which was explained by a Brazilian lawyer at an agricultural investment conference I attended in 2011 and in a subsequent interview, is for the Brazilian and foreign individuals to create a sort of limited partnership together in which one partner is a "hidden partner" (*socio oculto*) while the other is the "ostensible partner" (*socio ostensivo*). The ostensible partner, in this case the Brazilian, is the face of the partnership, while the hidden partner, in this case the foreigner, simply invests and receives returns in accordance with the partnership agreement between them. Whereas the aforementioned approaches take advantage of the oversimplicity of the "foreign" label, other strategies for getting around the restrictions are based on more general capital maneuvers. For instance, several research participants told me that it would be possible to acquire land via foreclosure on an unpaid mortgage. In this approach, a Brazilian landowner takes out a mortgage on his or her farm from a foreign lender. The Brazilian then promptly defaults on the payments, leaving the land in the hands of the foreign lender/buyer.

The Brazilian government's attempt at restricting foreign investment in farmland was ineffectual in large part because it addressed land only as national territory, not as the financialized commodity that it is increasingly becoming. Couched exclusively in terms of the foreign/domestic divide, the new restrictions are not good at recognizing the many ways in which capital flows between these categories, often passing through Brazilian entities on its way to farmland. If finance capital's interest in farmland markets continues to grow, regulations on "foreign" land acquisitions will likely become harder and harder to enforce.

One possible explanation for Brazil's misplaced focus on the foreigner is that the Brazilian government did not actually *intend* to put effective regulations in place. James Holston has called Brazilian land law "an instrument of calculated disorder" whose consistent inefficiency and indeterminacy ultimately serves to further land concentration.[71] Brazilian land economist Bastiaan Reydon, meanwhile, summarizes the historical relationship between elites, land, and the state in Brazil as follows: "The great ability of the Brazilian elite was always in creating rules that apparently aimed at the effective regulation of [land] markets. Both compliance with these rules and their circumvention, with the backing of the State primarily through a lack of enforcement, has always generated conditions conducive to speculation for the few."[72] Although the Lula and Rousseff administrations were deeply committed to smallholder agriculture on paper, their actions admit of doubt. The PT government's move to restrict foreign land acquisitions must be understood in the context of its other land-related actions, which include failing to enact comprehensive agrarian reform and reluctance to acknowledge the land claims of indigenous groups and communities descended from escaped slaves (*quilombos*).[73] Brazilian geographer Ariovaldo Oliveira went so far as to argue that the entire foreignization scandal was a "farce" fabricated by the national government to distract from its ongoing program of agrarian counterreform.[74] Whether or not the government's actions were quite this calculated, the reregulation certainly helped the Brazilian government chart a course between its conflicting commitments to both agrarian reform and large-scale industrial agriculture. Placing ownership restrictions on foreigners was a way for the government to score points with rural social movements by tackling upward price movements that could harm agrarian reform. But the simplistic foreign/domestic dichotomy on which the regulation was based also left substantial leeway for corporations to circumvent the law and go about their business.

## "Land of All of Us": The View from Western Bahia

To see how international agricultural investment is unfolding in Brazil, you cannot do better than to visit Western Bahia. Western Bahia is perfect for export agriculture: it is very flat, it has dependable rains, it has good highway infrastructure. Its level of agricultural development hits a sweet spot between the very well-established *cerrado* farming regions like Mato Grosso and the riskier frontier of *cerrado* agriculture in the neighboring states of Maranhão, Piauí, and Tocantins. As you enter Bahia state—whose state motto, incidentally, is "*terra de todos nós*" or "land of all of us"—from the west on highway BR-242 or southwest on BR-020, the first town to greet you is Luis Eduardo Magalhães, an agricultural boomtown that has grown out of nothing since the 1980s thanks to the *cerrado*'s transformation under Brazil's version of the Green Revolution.[75] Now travelers entering the municipality are met by a billboard reading "Welcome to Luis Eduardo Magalhães. The capital of agribusiness," and adorned with an advertisement for Ford trucks. The foreign presence here is almost palpable. L.E.M.—as it is affectionately known—is home to a soybean crushing facility owned by the multinational grain trader Bunge. Since the late 1980s, the region has also been home to migrant US farm families searching for cheaper land and better opportunities than those provided by the US Midwest.[76] But the international presence here goes beyond individual migrants. The highways of Western Bahia are spotted with signs marking farm entrances, often adorned with the logo of a foreign-owned agricultural operating company: Adecoagro, Agrifirma, BrasilAgro, Calyx Agro, Grupo Iowa, Multigrain, SLC Agrícola, Vanguarda.

Following highway BR-020 farther east, one arrives at Barreiras, a major agricultural city. When I visited Western Bahia with a research assistant in July 2012, the city of Barreiras was hosting its annual Barreiras Agricultural Exposition—"Expobarreiras." We wound our way through fairground rides, livestock pens, and displays of agricultural machinery to eventually arrive at the temporary headquarters of the Barreiras municipal government in a large, domed white tent. There we were immediately handed a maladroitly translated English-language brochure aimed at potential investors. "The production profile of the Cerrado is based on intensive and commercial agriculture," it proclaimed. "Large national and international groups of agribusiness companies are present here. A large-scale agricultural production with international quality standards has made of western Bahia the most promising agricultural frontier in the world." Both the brochure and a municipal official we spoke with stressed the continued availability of agricultural land for foreign investment. With 30 percent of the land still "open" and new railroads being constructed, the official concluded, "the future we see is very promising."

The MST leader for the Western Bahia region, whom we met later that day, painted a very different picture. Foreign investment, he said, fueled rapidly appreciating land prices, which in turn made it difficult for the government to expropriate farms and conduct its mission of land redistribution. There were over fourteen thousand families currently taking part in fifty-seven settlements and occupations across the region, he told us, yet the government had not expropriated any land since 2008. Infusions of foreign capital, he argued, also further the dominance of industrial agriculture in the region, which is linked to the displacement of rural families, soaring crime rates in cities like Barreiras, and an increase in cerebral aneurysms and other illnesses linked to agro-chemical poisoning. Not only had land reform ground to a halt; the landless were being poisoned by their industrial neighbors.

The one thing these accounts agree on is that international capital is reshaping the agricultural and social landscape of Western Bahia—and yet this is not reflected in the official records. A visit to a local *cartório* revealed that foreign land acquisitions were essentially unregulated. Within the first minutes of our interview, the person responsible for registering property ownership in this foreign-dominated region told us that he had *never* registered a property to a foreigner. Not during the decade of neoliberal deregulation, and not in the two years since reregulation had taken place. This is not to say that he had never registered properties that he knew to be owned by foreigners—that was a regular occurrence. Often, he said, the name of the farm was in another language, or the people registering it mentioned the American or Dutch owners. But none of this mattered, as long as it was occurring through a Brazilian company. "We have various companies that I know are of foreign capital, but . . . the documentation that they bring is all for a national company. . . . I have various registrations of these companies, but for me it is masked, right? It is disguised as national." Trying to determine whether a company is "really" Brazilian or not, he told us, is not a part of his job description, nor is it within his capacity. His obligation only begins when a company or individual declares itself to be foreign, which had yet to happen in his area. On paper, foreigners owned no land in his region.

The investor-backed land rush in Western Bahia, as in the rest of Brazil, bears traces of the country's long history of *grilagem*. For centuries, Brazilian elites have taken land first and worried about ensuring its legality later. As a result, there is perhaps nowhere that international investors could have found a set of domestic actors more willing and able to creatively circumvent landownership restrictions than in Brazil. But the land rush also builds on the legacy of *grilagem* more directly: the large properties that are so attractive to institutional investors were often amassed through extralegal means, including document fraud and violence. Investors who are drawn to Brazil for the potential of scale that it offers do not always adequately investigate the history of their farms.

In the Western Bahia municipality of Cotegipe, for instance, a 140,000 ha landholding now owned by Harvard University was originally acquired by local elites through *grilagem*. In the early 1990s, according to media and NGO reports, this land was still owned by the Brazilian government and was being legally settled by 240 families. These farmers built homes and cultivated crops with the intention of earning title to the land under Brazilian law. Instead, they were driven off by threats of violence from local elites; men arrived with guns, homes and crops were burned, and fences erected. The mega-property that resulted from this campaign of intimidation (140,000 ha is bigger than the city of Los Angeles) changed hands a couple of times before being sold to Caracol Agropecuária Ltd., a company backed by Harvard Management Company, which oversees Harvard University's $37 billion endowment.[77] TIAA has likewise been accused of purchasing farms in Maranhão and Piauí from companies associated with a notorious Brazilian land grabber, Euclides de Carli.[78]

In Brazil, as in the US (see chapter 1), the current transfer of farmland into the hands of investors builds on a foundation of historical dispossession. In Brazil, centuries of *grilagem* have furnished properties on the scale of operation desired by financial institutions, while a tradition of evading land law has provided the legal tools needed to acquire them. Yet the layers of complex subsidiaries and financing mechanisms involved in farmland investment have the effect of obscuring this bloodstained history.[79]

There was nothing unusual about Brazil's decision to restrict land acquisitions by foreigners. Many, or even most, countries have some sort of law regulating foreign landownership. These run the gamut from outright restrictions to simple reporting requirements.[80] In Indonesia, the law prevents foreigners from having freehold ownership over any land. Australia and Singapore have restrictions on the purchase of urban and residential land, respectively. Mexico restricts foreigners from owning land along border zones, while Peru restricts foreign landownership near military bases. Interestingly, limitations on foreign landownership that apply particularly to farmland are common. Morocco prohibits the sale of farmland to foreigners, and Turkey restricts rural landownership as a means to preserve village life. In Ireland, if control of a company with agricultural landholdings passes into foreign hands, the government must be informed within one month.[81] The Canadian provinces of Manitoba and Saskatchewan both limit farmland ownership by foreigners, as do several midwestern US states.[82] In response to the recent land rush, countries as diverse as Argentina, Hungary, and the Democratic Republic of the Congo also instituted limits or outright bans on farmland purchases by foreigners.[83]

Though laws against foreign landownership can have many motivations, including nationalism, racism, and xenophobia, the abundance of anti-foreign laws (or "alien land laws," as they are often called) focused particularly on farmland suggests that they can also serve to express a more general desire to keep landownership in the hands of those who farm it.[84] Common etymologies hint at this connection: alien landownership fuels fears about alienation from the means of production; *estrangeirização* contributes to an estrangement between land and people.

Restricting farmland acquisitions by foreigners may be one of the most politically feasible ways for governments to act in the face of growing demand for farmland from wealthy investors and corporations. I have argued above that this was the case in Brazil because of how broader debates about the social function of land have become boxed in by an overriding concern with agricultural productivity. Though Brazil's agrarian social movements adopt a much more expansive view of the moral economy of land,[85] the official legal playing field is circumscribed by the politics of productivity. In this political framework, productive use is seen as the only necessary condition to justify landownership, rendering broader imaginaries of a more socially just agricultural structure superfluous to the debate. While the politics of productivity has been enshrined in the Brazilian approach to land reform, it is by no means unique to Brazil; regulators everywhere are constrained to some degree by the idea—assiduously promoted by investors themselves, as discussed in chapter 2—that productive use is synonymous with beneficial agricultural investment.

Yet this approach to regulating land markets is deeply insufficient to address landownership by footloose finance capital. The clear-cut distinction between foreign and domestic has always been illusory; even in the 1970s, the Brazilian government had to set an arbitrary boundary line between the two categories at 50 percent foreign capital in order to make its restrictions on foreign land acquisitions functional. But as companies have become more transnational and finance capital more mobile, that distinction has grown even more tenuous. Where the land purchaser is a publicly traded transnational corporation with shareholders on five continents, a financial management company operating via holding companies out of an international tax haven, or a foreign pension fund partnering with a domestic operating company, "the foreigner" becomes a moving target, difficult to pin down for more than a moment. Today the foreigner provides an object for popular outrage and for government policy action, but it is an unstable signifier with limited ability to achieve a more equitable distribution of land.

# CONCLUSION

**What we call land is an element of nature inextricably interwoven with man's institutions. To isolate it and form a market for it was perhaps the weirdest of all the undertakings of our ancestors.**

—Karl Polanyi, *The Great Transformation*, 1944

Farmland is laden with varied and sometimes contradictory meanings: a commodity that was not produced and cannot circulate, a source of personal independence and of group identity, a productive asset that moonlights as a financial asset. Now it is increasingly attracting the attention of professional investors and being drawn into financial circuits. This book has explored the growing entanglements of finance and farmland by drawing out the slender filaments connecting financial change to agrarian change. These threads link the hard-won savings of retirees, unions, and universities to the financial professionals whose duty it is to make those savings grow. They link the daily decisions those professionals make about the assets in their care—decisions about risk and liquidity and the contours of an unknown future—to shifts in agricultural landownership in Illinois or Western Bahia. These threads are gossamer, spun from millions of digital financial transfers and thousands of investment pitches, but their effects are no less real.

Investor interest in farmland did not come out of nowhere in 2008; it was the result of cumulative changes, including the professional farm-management capacity acquired in the aftermath of past booms and busts, as well as institutional investors' ever-expanding search for new ways to deploy their swelling pools of capital. The food and financial crises simply pushed it over a tipping point, igniting unprecedented investor interest. Nor is farmland financialization easily reduced to a single, defining change. It can be seen in the growing numbers of farmland and agriculture-related investment funds, but also in a certain investor mind-set that treats land primarily as a source of financial income. It can be seen in the moral vision put forward by the investment community about its work, as well as in the emergence of new approaches for addressing agricultural risk and farmland illiquidity. Together, these developments in investment thought, strategy, and discourse are working to mold farmland into a desirable and accessible global asset class. In Polanyian terms, they attempt to further disembed land from the nonmarket institutions and noneconomic values with which it has long been associated.

Yet this disembedding remains an arduous and contingent process. To be successful, it must grapple with the peculiar idiosyncrasies and risks of agriculture—long a source of both opportunity and constraint for finance capital—as well as novel financial risks of its own making. It must contend, too, with moral friction stemming from the many species of value associated with agricultural land. Finally, it faces the threat of regulatory intervention on the part of the state, often tied to the politics of territorial sovereignty. These many obstacles notwithstanding, efforts to further disembed farmland from its social integument by transforming it into a global financial asset class could have profound consequences. These consequences merit serious scrutiny, as do nascent efforts to counteract them.

## Why the Financialization of Farmland Matters

That farmland is increasingly being incorporated into the multibillion-dollar portfolios of the wealthiest institutions matters for many reasons. It is just one facet of an ongoing accumulation of wealth at the top of the economic pyramid, but it is a particularly meaningful one because of the different significance land holds to different owners. While for institutional investors land is just so much financial ballast—a way to diversify away risk while making a respectable profit—for small farmers the world over, land is life. For a Malawian farmer, a single hectare of land could mean the difference between her children eating or not,[1] whereas institutional investors generally require hundreds or even thousands of hectares just to make a dent in their capital allocation to farmland. The stakes involved may be much larger for the investor, but they are infinitely higher for the farmer.

Financial-sector actors also have a set of motivations and obligations fundamentally different from those of other types of landowners, and this can affect the way that they treat their farms. Investor-owners may gesture toward their strong environmental or social values, but at the end of the day they will be judged on their economic performance. The growing tendency to evaluate companies on the "shareholder value" they produce is, in fact, reinforced by law; many types of financial and corporate managers have a legal obligation, known as a "fiduciary duty," to act in their clients' or shareholders' best interest by earning the highest possible rate of return. Fiduciary duty may discourage corporate managers from taking actions deemed socially or environmentally beneficial if doing so would in any way reduce the returns they earn for their clients.[2] Whereas landowner farmers take many things into consideration when making decisions about their land—landscape aesthetics, the affection they feel for the property, the good opinion of neighbors, the future they envision for their children—financial landowners are required to place economic returns above other forms of value.

At a 1980 US Senate hearing on pension-fund investment in farmland, Senator Gaylord Nelson worried aloud about the vastly different motives held by farmers and investors: "We will then replace the system of dispersed ownership of farmland by those who till the soil, and plan to hand it on to their children, with ownership by investors, speculators, and institutions interested in maximizing their economic gain. At some point, almost inevitably, this means selling the land to the highest bidder. No sentimental attachments to the land will keep the land in farming. No love of the soil will assure careful conservation for coming generations."[3] He went on to observe that any investment company that "allows sentiment to keep land in farming in the face of a good bid from an investor, would be violating its fiduciary responsibilities and would be subject to suit."[4] In 1980, pension-fund investment in farmland was still just a novel possibility; four decades later it is a pervasive reality, and his words continue to ring true.

Financial investment in land also changes the very meaning of ownership. A Brazilian operating company executive argued to me that increasing financial investment in land was leading to a "democratization of ownership." Speaking about his company's plans to go public, he argued: "When you list a company like ours, then you provide access to literally thousands of people that wouldn't have access to any kind of landholdings in the near future. . . . So this is how it ends, so to speak, the whole circle." His suggestion was that landownership has come full circle, from initially being very dispersed, to getting more and more concentrated in the hands of companies, and finally getting redistributed through the stock market. The CEO of Farmland Partners, the publicly traded farmland REIT, likewise argues that by giving stock market investors the chance to buy shares in US farms, his company is involved in the "democratization of real estate ownership."[5] By this logic, it could be argued that even pension funds are essentially redistributing farmland ownership among the millions of teachers, government employees, and steelworkers whose retirement savings they are using to amass agricultural real estate.[6]

Yet many people are left out of this model of financially mediated farmland ownership. Even in a wealthy country like the US, almost half the population has no money invested in the stock market—not in pension funds, not in mutual funds, not in 401(k) plans or individual retirement accounts.[7] In the Global South, meanwhile, landownership-via-finance will be relevant to a vanishingly small percentage of the population. The gaping wealth gap described by Thomas

Piketty will only widen as farmland all over the globe becomes a source of profit for wealthy and upper-middle-class investors in a small subset of rich countries.

What's more, financialized ownership—or "portfolio ownership," as Gerald Davis calls it—differs in crucial respects from traditional ownership. Coincidentally using farming as his example, Davis explains: "The term *ownership* evokes the family farmer working his ancestral land, patiently improving his patrimony for future generations. But portfolio ownership means something very different. Owning shares in a widely-held corporation merely gives a fractional claim on future residuals and pointedly excludes real control."[8] Portfolio ownership is ownership stripped of everything but the profits and losses. Gone are the responsibilities and relationships that ownership used to imply. And the spread of portfolio ownership is not limited to stock-market investing; Davis argues that, under financialization, we increasingly find ourselves in a "portfolio society" in which everything—including social relationships—is viewed as a form of capital that should produce returns: "To view stocks, bonds, education, jobs, friends, and neighborhoods as investments in a portfolio implies that they are, in some sense, a 'position,' not a commitment."[9] Absent commitment, landownership has little significance beyond the regular drumbeat of quarterly returns. Meanwhile, the people who actually farm the land will increasingly be tenants and will have none of the incentives of ownership to encourage careful stewardship of soil, water, or other resources.[10]

An increasing tendency to view land through the lens of "portfolio ownership," as Davis puts it, or as a "pure financial asset," in the words of David Harvey, will have very real effects on agriculture and the rural communities that depend on it. To the extent that institutional investors take relatively long-term positions in farmland markets—viewing land as a way to reduce portfolio risk through diversification over the long haul—their purchases may contribute to land price inflation in desirable agricultural regions. Their demand for large farms that can absorb their considerable capital also encourages farmland investment intermediaries to consolidate farms in order to increase their value. All of this will make it harder for farmers to expand their properties or get an initial toehold in landownership. Because appreciation is central to most farmland investment strategies, investors also shepherd their properties toward higher-value uses whenever possible, whether by transforming *cerrado* pastureland into a functioning farm or by transitioning a California vegetable farm to higher-value permanent crops such as almonds or wine grapes. In short: the financial sector's growing interest in agriculture is not scale neutral. By its very nature, it will tend to push agriculture toward larger-scale and more capital-intensive production.

This tendency to propel agricultural land toward large-scale, high-value operations will have ecological as well as social consequences. Industrial agriculture is a major contributor to climate change, producing greenhouse gas emissions through deforestation, fossil fuel–burning farm machinery, fertilizer use, and intensive livestock raising.[11] Small-scale farming based on agroecological production practices emits fewer greenhouse gases and may even absorb carbon, but this model of production is virtually incompatible with the logistical needs of investors.[12] Even in cases where appreciation is achieved through organic certification of farms, the higher land values that result will likely exert pressure on growers to keep land in production for as much of the year as possible and to produce only the most valuable cash crops rather than rotating in lower-value crops simply because they are good for the soil.[13] Investors frequently assert that their pursuit of land valorization precludes them from treating their properties poorly. This is probably true. But it also forecloses an entire world of agricultural possibilities that would replenish natural systems and foster social equity.

In the long run, as more and more investors treat farmland as a standard financial asset class, farmland markets may also become more volatile. At present, investors are drawn to farmland in part because its returns are relatively uncorrelated to the movements of the stock market, making it an exceptional portfolio diversifier. However, to the extent that land becomes an asset class like any other, its value will be increasingly linked to stock market yields and even more vulnerable (if possible) to interest rate changes.

## Possible Solutions

A decade since the financial sector's "discovery" of farmland, several civil-society-led approaches have arisen to curb that discovery's possible ill effects. Each can be seen as a very different approach to "re-embedding" farmland in nonmarket values and institutions. Three of these—corporate codes of conduct, activist campaigns targeting particular investors, and cooperative investment structures—have received particular attention. However, each of these approaches has shortcomings that call into question their potential to produce a genuine re-embedding of land.

### Corporate Codes of Conduct

The global land rush was quickly followed by a rush to produce guidelines on how the new farmland owners should go about acquiring and operating their properties. These go by various names—"codes of conduct," "guidelines," "principles," "frameworks"—but the underlying idea is the same: they outline rules of behavior that companies can choose to comply with in order to make their land deals more socially responsible and environmentally sustainable.

Some of these documents have been produced by intergovernmental organizations or forums with civil-society and private-sector participation. In 2010, the World Bank, the UN Food and Agriculture Organization (FAO), the UN Conference on Trade and Development, and the UN International Fund for Agricultural Development outlined seven "Principles for Responsible Agricultural Investment that Respects Rights, Livelihoods, and Resources" (PRAI).[14] These principles were endorsed by the international forum of governments known as the Group of 20 but scorned by many social-movement organizations, which argued that they were created in a top-down manner without civil-society input and that they simply served to legitimize large-scale agricultural investment projects.[15] Another prominent effort was sponsored by the Committee on World Food Security (CFS) of the FAO, which in May 2012 adopted "Voluntary Guidelines on the Responsible Governance of Tenure of Land, Fisheries and Forests in the Context of National Food Security" (VGGTs).[16] Many social movements saw these guidelines as more promising, because civil society organizations were heavily involved in their creation and because they take an explicitly human-rights-based approach. The CFS followed up this effort by embarking on a two-year consultation process with private-sector and civil-society organizations to create its own "Principles for Responsible Agricultural Investment in Agriculture and Food Systems" (rai—sometimes written lowercase to differentiate them from the World Bank-sponsored principles of a similar name), which were designed to be compatible with the VGGTs. These government-backed guidelines are the most prominent, but they have since been joined by a host of others put forth by the Group of 8, the US Agency for International Development, the Organisation for Economic Co-operation and Development, and the African Union, to name a few.[17]

Several voluntary codes of conduct have also emerged from the private sector. In 2011, a group of eight institutional investors convened under the auspices of the Principles for Responsible Investment (PRI), a membership organization for investors backed by the UN Environment Programme Finance Initiative and UN Global Compact. The five "PRI Farmland Guidelines" that resulted from this investor-led process are noticeably more vague and aspirational than those coming from the public sector, emphasizing investor *progress* toward more sustainable and ethical actions.[18] Large-scale land acquisitions can also fall under the guidelines produced by corporate-led initiatives relating to particular crops, such as the Roundtable on Sustainable Palm Oil, Roundtable on Sustainable Biofuels, and Round Table on Responsible Soy. Many companies also have their own in-house corporate codes of conduct, which they may or may not make public.[19]

As a means to address potential environmental and social harm caused by large-scale land acquisitions, voluntary guidelines fall short on a number of fronts. First, and most obviously, these guidelines are *voluntary*. Without strict enforcement mechanisms, companies can receive all the public relations benefits of signing on while doing little to actually comply with the guidelines. In this sense, they may allow investors to present a sanitized image while producing little change in their actual behavior. Jennifer Clapp argues that voluntary initiatives also fall short in a range of practical ways: they are not generally very successful at attracting signatories; the proliferation of competing initiatives is confusing and encourages laxer guidelines to attract participants; and for most companies the business case for implementing voluntary corporate social responsibility practices is pretty weak. In the case of financial investments in agriculture, Clapp argues, these general difficulties with voluntary codes are compounded: the complexity of the investments has the effect of "distancing" investors from outcomes; most investors do not have a recognizable brand name that requires such proactive efforts to avoid reputational risk; the short-term performance criteria on which financial manager performance is evaluated are not compatible with the long-term efforts required for sustainable agriculture.[20]

Additionally, as Saturnino Borras and Jennifer Franco observe, voluntary codes of conduct tend to fetishize certain mechanisms for achieving responsible investment without interrogating their actual usefulness. "Transparency," for instance, is treated as a good in and of itself, even though "transparency does not necessarily guarantee pro-poor outcomes" and "is not the same as accountability."[21] Likewise, voluntary guidelines pin major hopes on the concept of community participation. Prior consultations with community members, or even "partnerships" with local farmers or NGOs, are touted as a means to ensure that local people benefit from investments. These ideas are based on "a depoliticized and unrealistic vision of engagement between various actors . . . imagining equal footing and complementary interests where none exist."[22] By imagining away the power imbalance between investors and local people, such guidelines ignore the root cause of socially unjust outcomes.

Perhaps the most fundamental shortcoming of voluntary guidelines, however, is that they take the land deals as a given. They start from the premise that conversion to larger-scale, more capital-intensive agriculture is inevitable—or even necessary for rural development—and then strive to make those investments more environmentally and socially friendly. The two sets of guidelines developed under the auspices of the CFS are a partial exception to this rule. But even in that relatively egalitarian forum, civil society organizations had to fight tooth and nail to insert a clause suggesting that, in some cases, the most desirable outcome might be for a proposed land deal simply not to take place.[23] In short, codes of conduct, through their very conceptualization, cede ground in what should be a vigorous debate about what type of agriculture is best for food security, the environment, and rural livelihoods. Their underlying assumption that big, well-capitalized agriculture is best—or at least inevitable—led Olivier de Schutter, the former UN Special Rapporteur on the Right to Food, to condemn them as providing "a checklist of how to destroy the global peasantry responsibly."[24] To the investor they bring great benefits, sanitizing the project of any unwholesome connotations of "land grabbing" and thereby protecting shareholder returns from the potentially devastating effects of negative publicity. But for affected communities, they may just mean a kinder, gentler form of dispossession.

## Campaigns Targeting Investors

Some farmer advocates and environmental activists, for whom corporate codes of conduct sound too much like foxes guarding the henhouse, are instead attempting to use public pressure to force investors' hands. Corporate-focused campaigns have sometimes achieved considerable success by using negative publicity to shame corporations into behaving more responsibly (think of the anti-sweatshop-labor campaigns aimed at apparel companies like Nike and Gap). Today activists are using some of the same tactics to try to coerce farmland investors into changing their actions.

In April 2017, I joined a group of about twenty protesters outside TIAA's corporate headquarters on Third Avenue in Midtown Manhattan. The protest was led by Friends of the Earth US (FOE US) along with a coalition of other organizations, including ActionAid, GRAIN, the National Family Farms Coalition, and Brazil's Social Network for Justice and Human Rights (Rede Social de Justiça e Direitos Humanos).[25] The protesters brought with them several large boxes filled with petitions against TIAA's involvement in large-scale farmland acquisitions, as well as its stock-market investment in companies involved in deforestation for palm-oil production. We chanted ("When I say 'responsible,' you say 'not!' When I say 'land grabs,' you say 'stop!'") and waved signs. The action culminated in a farcical skit, in which one activist—dressed in a suit to represent a pension fund executive—ordered a large cardboard bulldozer with "TIAA" emblazoned on the blade to remove some farmers from their land. Mocking TIAA's ostensible commitment to responsible investing as a signatory of the UN PRI guidelines, he cautioned the bulldozer to remove the farmers from their land "gently." The dispossessed "farmers"—played by activists in straw hats—along with a person in an enormous orangutan costume, who represented biodiversity loss due to deforestation for palm-oil cultivation, shouted and pushed back until the bulldozer and executive slunk away in defeat.

This action aimed to make visible the connections between the middle-class Americans steadily turning over their savings to pension funds for management and the small farmers whose livelihoods and land access may be disrupted by farmland investment. With the company's corporate office building in the background, a silver-haired professor emerita and TIAA client shouted into a small bullhorn: "These investments of TIAA make my interests dependent on the stealing—let's call it what it is—stealing of land from poor peasant farmers." While corporations are only capable of thinking about short-term profits, she declared, "We're not corporations, we're people, who can think more broadly. We can think morally. What more basic moral truth is there than you shouldn't steal land from people who need it for survival?"[26] She and the other protesters aimed to wrench land back from the realm of financial logic by demanding that institutional investors take into account the social value of land.

Other speakers highlighted how TIAA's investments affect the land and people of particular places. "In Brazil there is no such thing as empty land," a Brazilian activist announced. A land purchase in much of the *cerrado*, she pointed out, means complicity with a long history of land grabbing: "The area where TIAA is in Brazil used to be public land, so the only way to acquire land in that region is by land grabbing, is by illegal land grabbing, by forging land titles." TIAA's investments in Brazil, she argued, meant an extension of commodity monocropping across the Brazilian landscape, destroying biodiversity and displacing peasant farmers and indigenous people. With speeches like these, the activists attempted to pin down the slippery movements of financial capital by tying them to particular concrete effects in specific geographic locations.

**FIGURE 5.1**    A retiree and TIAA account holder speaks at a protest against land grabbing at TIAA's Manhattan headquarters, April 20, 2017.
*Source:* Brandon Wu / ActionAid.

**FIGURE 5.2**  Activists performing a skit outside TIAA headquarters, April 20, 2017
*Source:* Brandon Wu / ActionAid.

The organizers of the campaign freely admit that TIAA is not the only, nor by any means the worst, actor when it comes to institutional investment in agricultural land. FOE US initially chose to target TIAA in its campaign against public-equity investments in palm oil. Then, in 2015, GRAIN and the Social Network for Justice and Human Rights released a report detailing TIAA's purchase of farmland from a wealthy Brazilian who stands accused of land fraud (*grilagem*) and violence against peasant farmers, and the organizations decided to join forces. In an interview, one of the campaign leaders explained the several reasons for TIAA's selection as campaign target. First, it is a household name and therefore susceptible to brand damage. Second, it works hard to project an image of being socially and environmentally responsible, leading the activists to conclude that it might be more willing to change than other, less conscientious financial institutions. Finally, as the organizer explained, "TIAA seemed like a really great first target because they represent relatively progressive beneficiaries—public-school teachers, university faculty, teachers, nurses, and so forth"—and these beneficiaries "are networked in theoretically very easy-to-mobilize ways." In other words, it is easier to target an institution whose clients include entire university faculty unions than to attempt to mobilize the many atomized investors in a mutual or hedge fund. The hope is to apply what the campaign organizer called "the domino theory of corporate campaigning," in which activists begin by getting one corporation to change its ways and then continue to pick harder and harder targets until they have transformed an entire sector.

The campaign has had some successes, including garnering one hundred thousand signatures on its petition asking TIAA to "go deforestation and land grab-free."[27] It has also encountered major challenges, however, which reveal the limitations of corporate campaigns. For one thing, TIAA has so far proven relatively immune to pressure. After a couple of initial meetings, the same campaign organizer explained, the company has essentially stopped engaging with the campaign. A cynical interpretation would be that the company has found it can more effectively manage reputational risk by releasing annual reports on "responsible investment in farmland" detailing its commitment to "effectively communicating with a broad set of stakeholders" than by actually engaging with those stakeholders.[28] A less cynical view would be that the company is already doing everything required to legitimize its landownership within the moral economy of the farmland investor. Though frustrated with TIAA's lack of responsiveness, the organizer acknowledged this possibility, saying, "I mean, I think we're coming from very different perspectives, right? Their mandate is to maximize profits for their shareholders. From certainly our side and the social movement side, it's to maximize equity for everybody, [and] not in the financialized meaning of the term 'equity.'" He concluded, "I think, within the logic of profit maximization, they probably do believe that they're doing the right thing."

Another major limitation of the corporate-campaign strategy is that it only works on companies with widely recognized brands. Many people have heard of TIAA, but how many people have heard of the private-equity companies Proterra Investments or NCH Capital, or the hedge fund Altima Partners? These companies all join TIAA on the list of fund managers that have raised the most capital for unlisted agriculture or farmland funds over the last decade (table 1.2). It is difficult (if not impossible) to put public pressure on an investment company that no one outside of Wall Street can name.

## Community-Based Farmland Investment

A final approach to re-embedding land markets in the social fabric lies in alternative ownership structures, such as community land trusts (CLTs) and real estate investment cooperatives (REICs). Unlike the previous two approaches, these are not explicit responses to the (financialized) land rush, but rather efforts to address the negative impacts that rentier landownership and real estate speculation, in general, can have on communities.

The CLT model is based on the idea that property ownership should not just benefit individual property owners but should instead serve the interests of the entire community, particularly its most disadvantaged members. Under the CLT model, the landowner is a private, nonprofit corporation that is governed by a board largely composed of local community members and homeowners/lessees of the trust. This entity owns the land, while the buildings on the land are available for purchase, and the building owners are granted extremely long-term, inheritable leases for the property upon which their buildings stand (ninety-nine years is a common lease term). The trust captures the land-value appreciation that stems from economic development of the surroundings—what John Stuart Mill termed the "unearned increment"—and can then redeploy those earnings for the benefit of the community. The CLT frequently also controls the appreciation of the buildings themselves through conditions embedded in the land lease, stipulating, for instance, that the trust retains the first option to purchase the building upon resale at a price equivalent to the owner's invested capital once adjusted for inflation and depreciation.[29] Through these mechanisms, CLTs remove any incentive to speculate in the properties they control and reduce the negative impacts of gentrification on their community. The CLT model echoes Henry George's calls for a single tax that would remove the entire value of property appreciation and use it for the social good, only it is generally implemented by community-controlled nongovernmental organizations rather than the state.[30]

CLTs have considerable potential for getting land into the hands of farmers. In fact, the CLT model was developed by black farmers and civil rights activists in the US South during the 1960s and 1970s in response to black land loss owing to discriminatory implementation of government farm programs, violent retaliation for civil rights activities, and the forced sale of properties that, in the absence of a written will, were passed down to many heirs (known as "heirs' property").[31] The first CLT in the US, named New Communities, was founded by black farmers Shirley and Charles Sherrod in 1969 on over 5,000 acres—the largest single tract of black-owned land at that time. Despite crushing economic burdens and USDA bias, this land trust has managed to persist and reinvent itself over time, now occupying a 1,600-acre former plantation.[32] Today, the majority of CLTs in the US are focused on residential housing, but the urban abandonment caused by the 2008 foreclosure crisis, combined with the popularity of the local-food movement, has recently led many CLTs to embrace urban agriculture as a means of neighborhood revitalization.[33] Other rural farmland-focused CLTs include the Sustainable Iowa Land Trust, based in Iowa, and the South of the Sound Farm Land Trust, based in Washington State.[34] In addition to CLTs, conservation land trusts and conservation easements that allow for agriculture can also be important tools for maintaining farmland as an incomplete commodity.[35]

Another alternative ownership structure aiming to democratize landownership is the REIC. REICs pool capital from community members and, after a democratic decision-making process, invest it in buying and refurbishing underutilized local real estate, often with a focus on commercial space. They generally have explicit aims of encouraging local business and keeping rents affordable, in addition to making modest returns for their community shareholders. Urban REICs, such as the New York City REIC and Minneapolis's NorthEast Investment Cooperative, show promise as a means to encourage the circulation of wealth within the local built environment.[36] One new REIC, however—Brooklyn-based Black Land Matters REIC—aims to combine investments in commercial, residential, and agricultural property.[37]

Though they hold considerable promise for subordinating finance to local democratic control, such efforts also face limitations. Rather than posing a radical challenge to the overall system of financialized landownership, they carve out small enclaves of sustainable real estate investment practice. In the best-case scenario, these enclaves may pave the way for a systemic overhaul by serving as models for reform; but in the meantime, they leave much unchanged. They also depend on—or at least are greatly facilitated by—a favorable regulatory environment, including government support for cooperative formation, tax credits that encourage investment in local business, and policies that increase the accessibility and flexibility of self-directed individual retirement accounts.[38] These challenges point to the importance of government involvement in regulating the financialization of farmland.

## Regulating Farmland Investment: Beyond the Politics of Productivity

Whereas neoliberal political discourse tends to position government and market action as antithetical to one another, in fact government regulation is essential to the smooth functioning of markets.[39] Intervention on the part of the state is, as Polanyi observed, central to both the commodification of land *and* to the implementation of protective regulations limiting the harm caused by that commodification. In other words, the state acts on both sides of the "double movement" toward and away from self-regulating markets.[40] As the most recent stage in the ongoing commodification of land, the financialization of farmland will inevitably be mediated *in some way* by state action. The question is simply how.[41] The governmental response to the land rush has so far run the gamut from no-holds-barred efforts to accommodate investors (via tax holidays, streamlined bureaucratic processes, and allocations of state-owned land, to name a few)[42] to measures aimed at slowing or stopping land acquisitions.

When national governments have responded with measures aimed at curbing the land rush, their response has frequently taken the form of limits or bans on land acquisitions by foreign entities.[43] This is likely because, after decades of neoliberal policy reform, many governments find their ability to intercede in land markets on economic or social grounds gutted, leaving state jurisdiction over national security as their primary recourse. Yet the enforcement problems experienced by Brazil show this type of measure to be an inadequate means of disciplining investment capital, which is now more mobile and less regulated than ever. Though restrictions on landownership by foreigners may be the most readily available and politically feasible way to regulate the land rush, they are in fact profoundly insufficient for addressing the challenges posed by financialized landownership. Though land remains fixed within national borders, investment capital does not; its national affiliation is murky, its movements elusive. The compatibility of such restrictions with xenophobic and nativist sentiments is an additional and powerful argument against focusing on foreigners.

The tendency of governments to address the land rush as a problem, first and foremost, of territorial sovereignty can also be partially attributed to the hegemony of a productivist vision of agricultural landownership. The long-standing Lockean idea that productive use justifies ownership has only been strengthened in recent decades by the narrowing scope of acceptable government action under neoliberalism and the perceived success of Green Revolution yield increases at fending off global hunger. This politics of productivity is a pillar upon which the emerging global farmland-investment community rests its moral case for landownership (as discussed in chapter 2). In Brazil, the politics of productivity has become legally entrenched in a constitutional definition of the social function of land that hinges on productive use—a definition that simultaneously aids the land-reform movement by justifying the occupation and expropriation of unproductive properties and constrains its scope of action when major land acquisitions are undertaken by investors with no intention of leaving land idle. This has contributed to whittling down the possible repertoire for state intervention in land markets (see chapter 4).

To effectively discipline the flows of finance capital into farmland markets, governments would have to embrace a much broader understanding of the social function of land—one that encompasses land's ability to produce wealth via appreciation. A century and a half ago, Henry George recognized that real estate speculation was a leading driver of economic inequality, yet his insights have long fallen out of favor in policy-making circles. George's work serves as a reminder of the power of tax policy to shape land markets by reducing incentives for speculation. His recommendation was to replace all existing taxes with a "single tax" that would completely absorb the unimproved value of land—the part of its value that humans had no role in creating. George argued that this land-value tax would obviate the necessity of land redistribution: "Let them continue to call it *their* land. Let them buy and sell, and bequeath and devise it. We may safely leave them the shell, if we take the kernel. *It is not necessary to confiscate land; it is only necessary to confiscate rent.*"[44] George's argument was that most taxes are accompanied by adverse incentives that reduce economic productivity; an income tax is a disincentive to work, a tax on manufactured goods is a disincentive to produce. In contrast, a tax on the unimproved value of land does not deter any beneficial economic activity; it does not translate into a reduced supply of land, nor does it deter landlords from making improvements to the land. Its primary effect is to drastically reduce the price of land. Even the economist Milton Friedman, strident opponent of all government intervention in the economy, gave his mild endorsement of this approach, calling it the "least bad tax."[45] Yet such a land-value tax has rarely been implemented because of the political difficulties to be overcome with current landowners. I submit that now might be a good time to take a hard second look at George's ideas.

There are other ways, too, in which tax policy could serve to curb the financialization of farmland. In the United States, the tax exemptions granted to pension funds, endowments, and some other institutional investors certainly play a role in encouraging the financialization of farmland. Meanwhile the policy that exempts REITs from taxation is essentially the exact opposite of George's land-value tax; in recent years it has actually served as a loophole allowing property-owning corporations—from private prisons to casino chains—to avoid paying federal property taxes by refashioning themselves as REITs.[46] Brazil, meanwhile, has a rural land tax, but its enforcement is spotty and ineffective; Brazil could greatly reduce the incentives for rural real estate speculation simply by uniformly enforcing this existing tax.[47] Governments can also embrace George's basic concept that the returns to the unimproved value of land should be used for the social good by supporting or even instigating the creation of CLTs, something which some municipalities are already doing.[48] These are just some of the simplest means by which national or local governments can intervene in land markets to reduce speculation.

The possible scope of government action in land markets is vast, running the gamut from full-fledged land nationalization or land redistribution to more limited measures designed to ensure that land serves a social function. This latter category can include restrictions focused on *landowner identity* that go well beyond the limited notion of the foreigner. Many countries, including Norway and Japan, as well as several US states, have laws on the books designed to keep land in the hands of family farmers while deterring corporate farming and rentier landownership.[49] It can also include restrictions on the *extent of landownership*. In 2010, at the same time as the Brazilian government was reinstating its ineffective restrictions on foreign landownership, a Brazilian coalition called the National Forum for Agrarian Reform and Justice in the Countryside (Forúm Nacional pela Reforma Agrária e Justiça no Campo) was carrying out a national "Campaign to Limit Land Ownership," which aimed to restrict the amount of agricultural land that could be owned by *anyone*, Brazilian or otherwise.[50] Though the campaign never had much chance of success, this kind of effort serves to broaden the national dialogue about how the social function of land can be achieved. Brazil also reminds us that if governments are to serve as effective guardians of the social function of land, they must be elected democratically and represent the people, not just elite or corporate interests.

Financial-sector demand for farmland should serve as a wake-up call. Land is too vital—to the pride of farmers, to the cultural identity of indigenous peoples, to the food security of rural communities—to be casually surrendered to wealthy institutions for whom it serves as just one more weapon in an investment arsenal. The profits it produces from rent and appreciation may be among its most tantalizing qualities, but they are also its least meaningful where social well-being is concerned. Land's increasing incorporation into financial circuits of capital presents an opportunity for a revived dialogue about how to define the contours of this incomplete commodity, a dialogue that should not shy away from the profound moral and political issues at stake. The investment decisions today being made in pension fund and endowment boardrooms will have repercussions long after the current boom subsides; how social movements and governments respond to those decisions will shape the future of the world's farmland.

# Notes

## INTRODUCTION

1. Jung-a, Oliver, and Burgis, "Daewoo to Cultivate Madagascar Land"; Berger, "Madagascar's New Leader"; Burgis, "Madagascar Scraps Daewoo."

2. GRAIN, "Seized"; *Economist*, "When Others Are Grabbing"; Zoomers, "Globalisation and the Foreignisation of Space." Throughout this book, I refer to a "farmland boom" or "global land rush," rather than using the language of a "global land grab." My reasons for this choice are three. First, I am looking at one small aspect of this much broader phenomenon—private, for-profit investment in farmland, with or without associated agricultural production. Second, the term "land grab" implies illegality. But such land acquisitions are often perfectly legal, either because host governments choose to make them so or because the investors are able to hire very good lawyers. In fact, this legality is often the crux of the problem. Finally, interviewing people with a wide range of stances on this topic, I attempted to use the same terminology when I spoke to everyone, rather than changing my language between talking to activists and investors. From an ethical standpoint, I simply do not feel comfortable referring to the vast majority of my research participants as "land grabbers" behind their backs.

3. Some land deals were of very questionable legality, such as when a US-based investment company was reported to have purchased 800,000 ha in South Sudan from the son of a tribal warlord (McConnell, "Secret Sale of Southern Sudan").

4. Burgis, "Great Land Rush." The Ethiopian government later revoked much of the Karuturi land concession, citing lack of progress in implementation, but in 2018 the company announced that it had resolved its disputes with the government and would continue the project on a smaller plot of 25,000 ha (Fikade, "Karuturi to Start Afresh").

5. On Russia and Ukraine: Kuns, Visser, and Wastfelt, "The Stock Market and the Steppe"; Visser and Spoor, "Land Grabbing in Post-Soviet Eurasia." On Laos and Indonesia: Kenney-Lazar, "Plantation Rubber"; McCarthy, "Processes of Inclusion and Adverse Incorporation." On South America: Borras, Franco, Gómez, et al. 2012, "Land Grabbing in Latin America."

6. Edelman, "Messy Hectares," 487; Deininger and Byerlee, *Rising Global Interest in Farmland*, 67; Nolte, Chamberlain, and Giger, *International Land Deals*, 8.

7. Nolte, Chamberlain, and Giger, *International Land Deals*, 7. The Land Matrix data only include deals initiated after 2000 and that encompass at least 200 ha (Nolte, Chamberlain, and Giger, 3). See Edelman, "Messy Hectares," for a discussion of the shortcomings of the Land Matrix and similar data sources on the land rush.

8. Malkin, "Thousands in Mexico City"; Lacey, "Across Globe, Empty Bellies Bring Rising Anger"; Knickmeyer, "In Egypt, Upper Crust Gets the Bread"; Food and Agriculture Organization of the United Nations, "World Food Situation."

9. McMichael, "Agrofuels in the Food Regime." For an excellent overview of the causes of the food crisis see Headey and Fan, "Anatomy of a Crisis."

10. Allan, "Virtual Water."

11. Fairhead, Leach, and Scoones, "Green Grabbing."

12. Wily, "'Law Is to Blame.'"

13. Minaya and Ourso, "U.S. Drought Shouldn't Scorch," 2; TIAA-CREF, "TIAA-CREF Announces $2 Billion"; TIAA-CREF, "TIAA-CREF Announces $3 Billion"; Nuveen, *Responsible Investment in Farmland*, 2018, 4; Nuveen, "By the Numbers."

14. Crippen, "CNBC Buffett Transcript"; O'Keefe, "Betting the Farm."

15. J. Wilson, "Corn Farms Are Hotter"; Mackintosh and Burgess, "Hedge Funds Muck In"; Bjorhus, "Hot Money"; O'Keefe, "Betting the Farm."

16. Cole, "New Black Gold"; Wasik, "More Precious Than Gold?"; J. Taylor, "Fields of Gold"; Gilbert, "Dirt Rich?"; Koven, "Exchange-Traded Farmland"; Land Commodities, *Agriculture and Farmland Investment Report.*

17. A 2010 survey of fifty-four farmland investment professionals asked respondents what they thought was driving end investor interest in farmland and agricultural infrastructure. Inflation hedging was the primary reason given, followed closely by farmland's low correlation to other asset classes. The "fundamentals" of rising agricultural prices due to food insecurity was a distant third (Highquest Partners, "Private Financial Sector Investment," 18).

18. For more on the relationship between urbanization, land speculation, and dispossession see Zoomers et al., "Rush for Land"; Levien, *Dispossession without Development.*

19. Epstein, "Introduction," 3. The value of "financialization" as a concept is up for debate. Brett Christophers ("Limits to Financialization") convincingly argues that—much like fellow terms "globalization" and "neoliberalization"—it is overused and limited in its analytical capacity. In relation to agricultural investment particularly, Stefan Ouma argues that financialization is a "historically blurry catch-all concept," making the case for its replacement with Sandro Mezzadra and Brett Neilson's ("Operations of Capital") more nuanced notion of "operations of capital" (Ouma, "Financialization to Operations of Capital," 83; see also Ouma, "Situating Global Finance," and Ouma, "Getting in between M and M'"). In using an overarching concept such as financialization, one must be careful not to become desensitized to the highly contextual ways in which finance reshapes different sectors in different places and at different times. I use the concept here, nonetheless, because I find it useful for drawing connections between developments in the world of agricultural investing and global-level political economic shifts of recent decades (Fairbairn, "Reinventing the Wheel?").

20. Neoliberalism is an ideological and policy movement that advocates reducing the role of government in the economy. There is something of a chicken-and-egg question regarding its relationship to financialization. For Duménil and Lévy ("Costs and Benefits of Neoliberalism," 17), "neoliberalism is the ideological expression of the reasserted power of finance." In other words, the consolidation and growing dominance of the financial class came first, followed by the neoliberal and globalizing policies that benefited their interests. For Kotz ("Financialization and Neoliberalism," 1), the opposite is true: "The immediate cause of the financialization process of recent decades is found in neoliberal restructuring, rather than financialization explaining the rise of neoliberalism." It is also important to note that the regulatory changes associated with financialization may not have been as ideologically coherent as they later appeared. Krippner (*Capitalizing on Crisis*) argues that financialization was the inadvertent product of a series of ad hoc government attempts to avoid crisis, not a concerted neoliberal policy drive.

21. Mazzucato, *Value of Everything*, xiv, 106–7.

22. Tomaskovic-Devey and Lin, "Income Dynamics," 543–45.

23. Stiglitz, "Capital Market Liberalization," 76.

24. G. Davis, *Managed by Markets*, 47–53; Fligstein, *Architecture of Markets*, 147–49; Froud et al., "Shareholder Value."

25. Espeland and Hirsch, "Ownership Changes"; Froud and Williams, "Private Equity and the Culture of Value Extraction."

26. Froud and Williams, "Private Equity and the Culture of Value Extraction"; Ho, *Liquidated*, 150–68; Lazonick and O'Sullivan, "Maximizing Shareholder Value."

27. "Security" is another word for tradable financial assets such as stocks and bonds.

28. Leyshon and Thrift, "Capitalization of Almost Everything"; *Economist*, "Bonds That Rock and Roll."

29. Greenberger, "Derivatives in the Crisis."

30. Krippner, *Capitalizing on Crisis*, 4. This understanding of financialization follows Arrighi (*Long Twentieth Century*) as well as Magdoff and Sweezy (*Stagnation and the Financial Explosion*).

31. Krippner, "Financialization of the American Economy"; Krippner, *Capitalizing on Crisis*. See also Orhangazi, *Financialization and the US Economy*.

32. Langley, *Everyday Life of Finance*; Martin, *Financialization of Daily Life*.

33. Books on this topic include Bjorkhaug, Magnan, and Lawrence, *Financialization of Agri-food Systems*; Clapp and Isakson, *Speculative Harvests*; Kaufman, *Bet the Farm*; Russi, *Hungry Capital*.

34. Clapp, *Food*; Clapp and Helleiner, "Troubled Futures?" The word "commodity" is used in two different ways. First, political economists frequently use it to refer to any good that was created for sale (see Polanyi, *Great Transformation*, 75). Likewise, they use the word "commodification" to refer to the process of turning into a commodity something that may not originally have been intended for sale. Second, in finance, the word "commodity" has a more particular meaning, referring to a class of commonly traded, minimally processed goods that are subject to uniform standards and measures, making them highly substitutable and leading to the emergence of a single global price. These traded commodities include metals (e.g., gold, copper, tin), energy commodities (e.g., crude oil, natural gas, ethanol), livestock and meat (e.g., feeder cattle, pork bellies), and other agricultural commodities (e.g., soybeans, coffee).

35. Masters, "Testimony"; Masters and White, "Accidental Hunt Brothers"; Wahl, "Food Speculation"; Wray, "Commodities Market Bubble." Others reject this explanation, attributing the price rise to a simple case of not enough supply and too much demand (Bobenrieth and Wright, "Food Price Crisis"; Irwin and Sanders, "Index Funds").

36. Burch and Lawrence, "Towards a Third Food Regime"; Avis, "Private Equity Money and Grocery Retailers."

37. Isakson, "Food and Finance," 6.

38. Murphy, Burch, and Clapp, *Cereal Secrets*, 27–29.

39. Even agriculture-sector corporations seemed eager, until recently, to distance themselves from actual agricultural production. In the 1990s scholarly observers noted a growing trend toward contract farming models, in which the corporation controls upstream and downstream activities while small farmers are left bearing the economic and natural risks of raising the livestock or growing the crops (Watts, "Life under Contract").

40. Commentators sometimes distinguish between the "real economy" and the "financial economy." In reality, however, the two are not fully separable (Magdoff and Sweezy, *Stagnation and the Financial Explosion*, 94). Financial innovation always builds on the "bread-and-butter of income flows from real assets" (Leyshon and Thrift, "Capitalization of Almost Everything"), which are rooted in material substance and geographic place (Pike and Pollard, "Economic Geographies of Financialization").

41. FitzSimmons, "New Industrial Agriculture"; Goodman, Sorj, and Wilkinson, *From Farming to Biotechnology*; Kautsky, *Agrarian Question*; Lewontin and Berlan, "Technology, Research, and the Penetration of Capital"; Mann and Dickinson, "Obstacles."

42. Polanyi, *Great Transformation*, 76. See Christophers, "For Real," for a critique of the notion of "fictitious commodities." See note 34 above for different meanings of the word "commodity."

43. Polanyi, *Great Transformation*, 60, 79.

44. Espeland, "Value-Matters," 1843–44; Shipton, *Mortgaging the Ancestors*.

45. D. Hall, *Land*, 11–14.

46. Li, "What Is Land?," 592; Visser, "Running Out of Farmland?," 185; Ducastel and Anseeuw, "Agriculture as an Asset Class," 200.

47. Ouma, "From Financialization to Operations of Capital," 82.

48. Desmarais et al., "Investor Ownership"; Ducastel and Anseeuw, "Agriculture as an Asset Class"; Kish and Fairbairn, "Investing for Profit"; Kuns, Visser, and Wastfelt, "Stock Market and the Steppe"; Larder, Sippel, and Lawrence, "Food Security Narratives"; Li, "What Is Land?," "Rendering Land Investible," and "Transnational Farmland Investment"; Magnan, "Financialization of Agri-Food"; Ouma, "Financialization to Operations of Capital," "Situating Global Finance," and "This Can('t) Be an Asset Class"; Sippel, "Food Security or Commercial Business?" and "Financialising Farming"; Sippel, Larder, and Lawrence, "Grounding the Financialization of Farmland"; Sommerville and Magnan, "Pinstripes on the Prairies"; Visser, "Running Out of Farmland?"; Williams, "Feeding Finance."

The first draft of this manuscript was completed in 2014 when the literature on finance-sector investment in farmland was still very limited. I was influenced primarily by Andrew Gunnoe's pathbreaking work on the financialization of timberland in the US (Gunnoe and Gellert, "Financialization") and by an early version of what would become Tania Li's article "What Is Land?" However, in the four years it took me to transform that earlier work into a manuscript worthy of submission (an embarrassingly slow pace, which I can only excuse by pointing to the two babies I produced in the same period), many excellent articles were published making arguments similar to my own. In particular, the work of André Magnan and colleagues on farmland investment in Canada, Sarah Sippel and colleagues in Australia, Oane Visser and colleagues in Ukraine and Russia, Antoine Ducastel and Ward Anseeuw in South Africa, as well as Stefan Ouma's overviews of the sector, are all empirically rich and theoretically insightful. I cite these related publications throughout and strongly encourage reading them in addition to (or instead of) this book.

49. HighQuest Partners, "Private Financial Sector Investment," 1; Wheaton and Kiernan, "Farmland," 5. HighQuest Partners asked market participants to estimate the size of the farmland and agricultural-infrastructure asset class. Wheaton and Kiernan's analysis was based on the authors' estimates of farmland owned by institutional investors specifically. There is some continuity between these two studies, however, in that the Wheaton and Kiernan article was coauthored by a HighQuest Partners researcher.

50. Willis Towers Watson, "Global Pension Assets Study," 3.

51. Jacobius, "More Investors Turn to Farmland"; Nuveen, "By the Numbers."

52. Zhang, "Farmland Value Survey Iowa"; *180 Graus*, "Preço da terra"; Savills Research, "International Farmland Focus 2014," 5.

53. Aside from having a lot of capital at their disposal, there is also reason to believe that investors may behave differently from other farmland market participants. In their research on farmland investment in Saskatchewan, Magnan and Sunley ("Farmland Investment and Financialization") found that investors paid more for farmland, on average, than other market participants, and that some bought and resold large tracts of land within short spans of time, suggesting that they were "flipping" it for speculative profit.

54. Piketty, *Capital in the Twenty-First Century*.

55. A recent study of investment returns for sixteen advanced economies from 1870 to 2015 found that investment in *residential* real estate has historically been a better way to generate income from wealth even than investment in the stock market—producing similar rates of return but with less volatility (Jordà et al., "Rate of Return on Everything").

56. Nader, "Up the Anthropologist," 5.

57. Abolafia, *Making Markets*; Ho, *Liquidated*; Zaloom, *Out of the Pits*.

58. My approach drew inspiration from "multi-sited ethnography" (Marcus, "Ethnography in/of the World System") and from "global commodity chain analysis" (Gereffi, "Buyer-Driven Global Commodity Chains"; Bair, "Global Capitalism and Commodity Chains"). However, rather than ethnography, I used a combination of other qualitative research methods: in-depth interviews, participant observation, and document analysis. And because farmland does not circulate like other commodities and therefore cannot be traced through the stages of production and trade, I instead interviewed the network of actors involved in facilitating farmland purchases, as well as the regulators and activists attempting to restrict them.

59. Brenner and Theodore's concept of "actually existing neoliberalism" is a useful reference point for thinking about how global processes are instantiated in particular national contexts (Brenner and Theodore, "Cities and the Geographies").

60. Most interviews were recorded, with participant consent, to later be transcribed and coded for key themes; if participants were uncomfortable with being recorded, I took detailed notes instead.

61. The reality is more complicated than these simple categories. Some companies, for instance, fall into more than one category at once. Such is the case with TIAA, the pension fund discussed above, which began by investing its own capital into farmland but then branched out into creating farmland funds for third-party investors, making it simultaneously an investor and an asset manager. To make matters more semantically complicated, most institutional investors are actually asset managers from the start. For instance, a pension fund is a type of institutional investor, but the real end investors are the myriad schoolteachers, firemen, and bus drivers, whose retirement savings the pension fund is charged with managing. To keep things simple, I will refer to these entities only as "investors." It is also important to note that the connections between the actors I interviewed were not linear. Though there were instances in which I talked to people in different countries about the same investment project, my goal was to understand the farmland investment sector as a whole, not to trace individual land deals from start to finish.

62. Garud, "Conferences as Venues"; Leivestad and Nyqvist, *Ethnographies of Conferences.*

63. Because most of my interviews took place in the US and Brazil, the majority of interview participants had at least some investments in North or South America. However, many had investments on more than one continent. Interviewees included seven individuals involved in farmland investment in Eastern Europe and the former Soviet Union, five in Africa, and five in Australia. I also spent two months researching farmland investment in Mozambique during 2010, which, though not discussed directly in this book, contributed to my overall understanding of the land rush (see Fairbairn, "Indirect Dispossession"). The investment conferences I attended were mostly framed as global events, but, with the exception of one I attended in the Middle East, they nonetheless primarily attracted participants from Western countries—particularly Europe, the Americas, and Australasia. It is important to remember that the cultures of finance are many, and my findings would likely have been quite different had I instead focused on, for instance, Islamic agricultural investors.

64. Magnan ("Financialization of Agri-Food") reveals the importance of studying the financialization of agriculture within particular national contexts through a comparison of Canada and Australia. Sippel, Larder, and Lawrence ("Grounding the Financialization of Farmland") further contextualize finance-backed investments at the local level through an examination of their reception by communities in rural Australia.

65. OECD, "Pension Funds in Figures," 1; McGrath, "New York, the Hedge Fund Capital."

66. Preqin, "Special Report: Agriculture," 5.

67. The first four countries were Indonesia, Ukraine, Russia, and Papua New Guinea. The Land Matrix database contained forty-five large-scale land deals in Brazil since

2000, encompassing over 2 million ha of land (Nolte, Chamberlain, and Giger, *International Land Deals*, 17–18). See note 7 above for more on the Land Matrix dataset and its limitations.

68. Cardoso and Faletto, *Dependency and Development*; Evans, *Dependent Development*; Frank, *Capitalism and Underdevelopment*.

69. I let my Brazilian interview participants decide whether the interview should take place in my (flawed but sufficient) Portuguese or their (sometimes quite impressive) English. The result was that my interviews with public-sector and civil-society actors were mostly in Portuguese, whereas my interviews with private-sector actors—many of whom had advanced degrees from English-speaking countries and used English regularly to speak with investors—were mostly in English. The interviews were all transcribed in the language in which they took place, and then I translated any quotes from Portuguese-language interviews that appear in this book. I also made the difficult choice to lightly edit the grammar in quotations from English-language interviews with non-native speakers. The people I was interviewing were successful and often very articulate professionals, but sometimes I read over the transcripts and was surprised to find them riddled with grammatical errors that made the speaker sound tongue-tied or even a little buffoonish. I felt that a denaturalized transcript, though it perhaps loses something of its immediacy, was more respectful to interview participants as well as conveying a more accurate picture of my experience of these conversations. Both the participants and I were accustomed to making do with the language resources available to us, brushing past them to get to the interesting and complex topic at hand. If they switched haphazardly between past and present tenses or struggled to find the correct word at times, this was not a representation of their expertise and, I felt, would only distract from the content of what they were actually saying (for thoughts on such transcription choices see Oliver, Serovich, and Mason, "Constraints and Opportunities," 1282).

70. I did not accept all such quid pro quos. Early on, I was offered free admission to an agricultural investment conference in exchange for the attendee lists from two competing conferences I had attended. These lists—which include the names of prospective investors—are one of the perks of conference attendance. I regretfully declined to provide the information and was not admitted to the conference.

71. Babb, "Sociologist among Economists," 50. Babb points out that this dynamic is not universal. It certainly applies in the US and Brazil, however. I mostly went with the flow of these gendered interactions. I bought a skirt suit, wore makeup, and generally attempted to emulate the kind of professional femininity I saw performed by the (relatively few) women in investment spaces. When interviews took place over coffee or a meal, I always attempted to pay the bill but never succeeded.

72. For more thoughts on gendered interview dynamics see Arendell, "Researcher-Researched Relationship," and Pini, "Interviewing Men."

## 1. FARMLAND INVESTMENT COMES OF AGE

1. Magnan, "Financialization of Agri-Food"; Sippel, "Financialising Farming"; Sippel, "Food Security or Commercial Business?"; Sommerville and Magnan, "'Pinstripes on the Prairies.'"

2. For a more complete picture of the long-standing entanglements between agriculture and finance in North America (including, particularly, the role of government in agricultural finance, which is barely discussed in this chapter) see Martin and Clapp, "Finance for Agriculture"; Ouma, "Financialization to Operations of Capital"; and Williams, "Feeding Finance."

3. Similar processes were at work in other countries; Great Britain, for instance, also saw an increase in financial landownership beginning in the 1970s (Massey and Catalano, *Capital and Land*, 114-138).

4. Native Land Digital, "Native Land."

5. For more on the relationship between financialization of land and dispossession of indigenous peoples see Sommerville and Magnan, "'Pinstripes on the Prairies,'" and Ekers, "Financiers in the Forests."

6. Peoples et al., *Anatomy of an American Agricultural Credit Crisis*, 98–99; Stam, Koenig, and Wallace, *Life Insurance Company Mortgage Lending*, 5, 8.

7. Stam, Koenig, and Wallace, *Life Insurance Company Mortgage Lending*, 11.

8. Notes on figure 1.1: Loan data do not include farm dwellings. Government category combines loans from various sources, including the Farm Credit System, the Farm Service Agency (which was previously the Farm Security Administration and then the Farmers Home Administration), and Farmer Mac. It does not include storage facility loans. Nominal prices were converted to 2016 dollars using the Consumer Price Index (CPI-U).

9. For overviews of US farmland market dynamics during the twentieth century see Henderson, Gloy, and Boehlje, "Agriculture's Boom-Bust Cycles"; Lindert, "Long-Run Trends." Though I focus here on the national level, land markets have distinct regional and local dynamics, meaning that land values may be booming in some areas while they are stagnant in others.

10. H. T. Johnson, "Postwar Optimism"; Saloutos and Hicks, *Agricultural Discontent*, 87–99; Shideler, *Farm Crisis*, 38–39. The relationship between interest rates and farmland values is discussed in detail in chapter 2.

11. Saloutos and Hicks, *Agricultural Discontent*, 100–105; Shideler, *Farm Crisis*, 46–58.

12. Notes on figure 1.2: Agricultural land values include buildings. Data from 1880 to 1910 are per decade; data from 1910 to 1916 are annual. Nominal prices converted to 2016 dollars using the Consumer Price Index (CPI-U).

13. Fitzgerald, *Every Farm a Factory*, 111.

14. Fitzgerald, 113.

15. *Los Angeles Times*, "New Farming Job Created"; Potter, "Farm Managers"; *Wall Street Journal*, "Chain Farm Method." See also Scott, *Seeing Like a State*, 198–99.

16. H. T. Johnson, "Postwar Optimism"; Woodruff, *Farm Mortgage Loans*, 7–36. Woodruff describes a variety of structural factors that led the life insurance companies to make unsound farm loans during the late 1910s. One factor was that their capital under management was building up, and they needed investment outlets. Another was that there was increasing competition between private mortgage providers and the newly established Federal Land Banks, leading to a relaxation of loan criteria. Additionally, the long distances between East Coast lenders and western farmers meant that most insurance companies made loans through a system of "loan correspondents" whose salary was based on commission, leading them to emphasize loan quantity over quality. The easy credit available at that time—much of which was provided by the federal government—contributed greatly to the farmland bubble that inflated throughout the 1910s.

17. Fitzgerald, *Every Farm a Factory*, 113.

18. Fitzgerald, 112.

19. Stam, Koenig, and Wallace, *Life Insurance Company Mortgage Lending*, 9.

20. Barnett, "US Farm Financial Crisis"; Peoples et al., *Anatomy of an American Agricultural Credit Crisis*, 6–29; Friedmann, "Political Economy of Food," 40.

21. Peoples et al., *Anatomy of an American Agricultural Credit Crisis*, 9, 20.

22. Barkema, "Farmland Values," 19.

23. Flint, "Solid Gold"; Crittenden, "Farmland Lures Institutions." Not all life insurance companies responded in the same way to the 1970s boom. In fact, some took advantage of high farmland prices to sell farms they had acquired during the foreclosures of the 1920s and 1930s (Peoples et al., *Anatomy of an American Agricultural Credit Crisis*, 14).

24. Barnett, "US Farm Financial Crisis"; Henderson, Gloy, and Boehlje, "Agriculture's Boom-Bust Cycles," 88; Peoples et al., *Anatomy of an American Agricultural Credit Crisis*, 6–29.

25. Hoppe, *Structure and Finances*, 7; Daniel, *Dispossession*.

26. Dudley, *Debt and Dispossession*.

27. Stam, Koenig, and Wallace, *Life Insurance Company Mortgage Lending*, 33; Peoples et al., *Anatomy of an American Agricultural Credit Crisis*, 39.

28. Stam, Koenig, and Wallace, *Life Insurance Company Mortgage Lending*, 36. See also Senf, "Life Insurance Company Ownership."

29. Stam, Koenig, and Wallace, *Life Insurance Company Mortgage Lending*, 37.

30. Stam, Koenig, and Wallace, 37.

31. Schneider, "As More Family Farms Fail."

32. HighQuest Partners and Koeninger, "History of Institutional Farmland Investment," 3.

33. Li ("Rendering Land Investible," 589) similarly notes the path dependence of land booms and busts. This is among several other interesting observations on the temporality of "rendering land investible."

34. Bleiberg, "Country Slickers."

35. *Forbes*, "Land Anyone?"

36. US Congress, "Ag-Land Trust Proposal," 185, original emphasis.

37. Phelps, "Corporate Farming Statutes." There had been prior state legislative efforts to combat corporate farming (Harl, "Farm Corporations"), though the 1970s saw the biggest wave of restrictive laws. Nine states still have such laws in place today (National Agricultural Law Center, "Corporate Farming Laws").

38. Quoted in Phelps, "Corporate Farming Statutes," 447–48.

39. DeBraal and Krause, "Corporate, Foreign, and Financial Investors"; Phelps, "Corporate Farming Statutes."

40. US Congress, "Ag-Land Trust Proposal," 3.

41. US Congress, 44.

42. US Congress, 45.

43. US Congress, 34.

44. Martin and Clapp ("Finance for Agriculture") point out that the state has always played an active role in mediating the relationship between agriculture and finance, though the nature of that mediation has changed over time.

45. US Congress, "Ag-Land Trust Proposal," 32.

46. This idea is particularly associated with Thomas Jefferson, who believed that landownership by independent, yeoman farmers was the key to American democracy, culture, and virtue. The landowning small farmer, he thought, was America's moral core—beholden to no one, the landowning farmer could act and vote his conscience. In a letter to James Madison, Jefferson made the case for agrarian democracy: "The earth is given as a common stock for man to labor and live on . . . it is not too soon to provide by every possible means that as few as possible shall be without a little portion of land. The small landholders are the most precious part of a state" (Letter from Thomas Jefferson to James Madison, October 28, 1785). This ideal lay behind the Homestead Act of 1862 and other early government efforts at promoting widely distributed landownership.

47. US Congress, "Ag-Land Trust Proposal," 40.

48. US Congress, 36.

49. US Congress, 70.

50. Clark, "Pension Fund Capitalism."

51. Clapp and Helleiner, "Troubled Futures?," 186. Position limits on noncommercial traders were put in place by the Commodity Exchange Act of 1936.

52. Clapp, *Food*, 140. A commodity index is similar to the famous Dow Jones Industrial Average, but rather than company stock, it measures the performance of a basket of agricultural and nonagricultural commodity derivatives. Commodity index *funds*, in turn, constitute a mixture of commodity derivatives tailored to track the performance of a particular commodity index. Investors buy into an index fund directly through a bank, allowing them to get exposure to commodity futures markets without the knowledge or hands-on participation that this would generally require.

53. This regulatory loosening was cemented by the Commodity Futures Modernization Act of 2000, which confirmed that OTC products were exempt from regulation (Clapp and Helleiner, "Troubled Futures?," 187).

54. This entire discussion is based on the work of Jennifer Clapp (Clapp and Helleiner, "Troubled Futures?"; Clapp, *Food*; Clapp and Isakson, *Speculative Harvests*).

55. Magdoff and Sweezy, *Stagnation and the Financial Explosion*; Arrighi, *Long Twentieth Century*. See also Harvey, *Enigma of Capital*. These thinkers agree that a crisis of over-accumulation was the ultimate cause of declining corporate profits in the 1970s but disagree on the proximate cause. Arrighi and Harvey largely blame increasing international corporate competition, while Magdoff and Sweezy blame growing corporate concentration and the rise of "monopoly capital." They also differ in how they view financialization historically. For Harvey, financialization is closely tied to the political and ideological rise of neoliberalism in the latter third of the twentieth century. Arrighi, on the other hand, sees the shift to financial channels of accumulation as a historically recurring process. He describes several historical "cycles of accumulation," each characterized by a period of material expansion, followed by heightened competition and stagnating profits, and finally a period of financial expansion in which firms switch from commodity production and trade to financial activities. Each cycle was organized under a different global hegemon; the four cycles of accumulation he discusses were led by the Genoese city states (fifteenth to early seventeenth century), the Dutch provinces (seventeenth to late eighteenth century), Britain (the late eighteenth century to early twentieth century), and the United States (the early twentieth century to the early twenty-first century). The US-led cycle of accumulation that occurred in the twentieth century, he argues, shifted into a phase of financial expansion in the early 1970s. The US government, working to maintain its hegemony, facilitated this shift through the abandonment of gold convertibility for floating exchange rates, the adoption of tight monetary policy and high interest rates, and deregulation of the banking sector.

56. See Harvey, "Urban Process under Capitalism," for a discussion of "capital switching" from commodity production into the built environment, and Harvey, "Globalization and the Spatial Fix," for a discussion of the spatial fix with an emphasis on geographic expansion.

57. Gunnoe and Gellert, "Financialization," 272. See also Gunnoe, "Political Economy of Institutional Landownership"; Gunnoe, "Financialization of the US Forest Products Industry"; Binkley, Raper, and Washburn, "Institutional Ownership of US Timberland"; Zhang, Butler, and Nagubadi, "Institutional Timberland Ownership"; Ekers, "Financiers in the Forests."

58. Conrad, "Farmland Asset Class Evolution."

59. Gunnoe and Gellert, "Financialization," 269.

60. In 1980, with the farmland boom still ongoing, a new company called American Agricultural Investment Management Company (AAIMC) was founded by three former

executives at Northern Bank and Trust Company of Chicago. Like Ag-Land Trust, AAIMC's primary purpose was to buy farmland for pension funds and then lease it to tenant farmers, but while Ag-Land Trust would have been a single fund, AAIMC proposed to set up separate accounts for each client. A pension fund would put $5–10 million under the management of AAIMC, whose managers would use this capital to create a geographically diversified portfolio of five to seven farms for the client (Orr, "Pension-Fund Farm Ventures"). The AAIMC founders hoped that their proposal was different enough that they could avoid the firestorm created by the Ag-Land Trust proposal, but they were destined for disappointment. Congressional hearings were called once again (this time under the auspices of the Senate Select Committee on Small Business), and the General Accounting Office issued a study on pension fund investment in farmland (US Congress, "Investment of Pension Funds in Farmland"; US General Accounting Office, "Pension Fund Investment in Agricultural Land"). However, the founders of AAIMC persevered, ultimately raising $16 million from two retirement fund clients. Another fund, Growth Farm Investors, was founded in 1981 but went unsubscribed after the crisis hit (HighQuest Partners and Koeninger, "History of Institutional Farmland Investment," 2).

61. The government response was also notably different. Martin and Clapp ("Finance for Agriculture," 550) argue that the nature of the relationship between agriculture, finance, and the state has changed over time: "Whereas the state has taken explicit measures to ensure that agriculture was supported by finance at various times, more recently states have instead ensured that financial markets were supported by agriculture."

62. There were others as well, but these were by far the biggest. This number was winnowed down from roughly eight at the end of the 1980s: Batterymarch AgriVest, Cozad/Westchester, Equitable Agri-Business, John Hancock Life, Metropolitan Life, Morgan Stanley, Phoenix Mutual Life, and Prudential Life (Fritz, "Institutional Investment in US Farmland"; HighQuest Partners and Koeninger, "History of Institutional Farmland Investment," 5).

63. Though Prudential's TIMO was ultimately acquired by Hancock Timber Resources Group (Zhang, Butler, and Nagubadi, "Institutional Timberland Ownership," 357).

64. Wise, *Investing in Farmland*; Wise, *Farmland Investment Strategy*.

65. HighQuest Partners and Koeninger, "History of Institutional Farmland Investment," 5; author interview with Phil, manager at US-based farmland management firm.

66. In recent years, this core group of asset managers has been joined by others with a similar profile, such as Halderman Real Asset Management, which began offering separate accounts in US farmland to institutional investors in 2013 and was renamed US Agriculture after a 2016 merger (US Agriculture, "US Agriculture and Halderman Real Asset Management Announce Merger").

67. Stiglitz, *Roaring Nineties*.

68. Conrad, "Farmland Asset Class Evolution."

69. Citing member confidentiality, a NCREIF representative could not tell me whether there were any particular bumps in index membership during this period. However, he agreed that it would be accurate to see this trend as indicative of industry growth as mediated through index contributor growth.

70. National Council of Real Estate Investment Fiduciaries, "Farmland Property Index."

71. Orange County Employees Retirement System, "Investment Manager Monitoring Subcommittee Meeting"; Fritz, "Cash Backlog Builds."

72. Fairbairn, "Indirect Dispossession."

73. Several recent papers have pointed out the need to tease apart the category of "finance capital," revealing the different motivations, approaches, and receptions of the various types of actor it contains (Knuth, "Global Finance and the Land Grab"; Ouma,

"From Financialization to Operations of Capital" and "Situating Global Finance"; Sippel, "Financialising Farming"; Williams, "Feeding Finance").

74. Knuth, "Global Finance and the Land Grab."

75. As Magnan ("Financialization of Agri-Food," 8) observes, however, not all investors agree that leasing out the land is the less risky approach.

76. Meyer, "Great Land Rush."

77. Payne, "Case for African Agriculture."

78. Meyer, "Great Land Rush."

## 2. FARMLAND VALUES

1. Zero Hedge, "Is TIAA-CREF Investing."

2. The word *physiocracy* derives from the Greek words *phýsis* (nature) and *krátos* (power) (Vaggi, "Physiocrats").

3. Gómez-Baggethun et al., "History of Ecosystem Services"; Hubacek and van den Bergh, "'Land' in Economic Theory."

4. Mazzucato, *Value of Everything*, 9.

5. Smith, *Wealth of Nations*, vol. 1, chap. 11, para. 262.

6. Mazzucato, *Value of Everything*, 34–40.

7. Malthus, *Essay on the Principle of Population*. Malthus was one of the few classical economic thinkers with anything good to say about landlords; for him, rent constituted a sort of bonus wealth that went to the landlord in return for judicious resource management (Heilbroner, *Worldly Philosophers*, 98).

8. Ricardo, *Principles of Political Economy*, chap. 2, para. 15.

9. Ricardo, chap. 2, para. 1–6. The theory of "differential rent" contained insights that would later be generalized to many other areas of economic life by the so-called marginal revolution in economics.

10. Heilbroner, *Worldly Philosophers*, 95; Foley, *Adam's Fallacy*, 77.

11. Marx, *Capital*, vol. 3, pt. 6, chap. 37. There is much, much more to say about Ricardian and Marxian theories of rent. For a good overview see Ward and Aalbers, "'Shitty Rent Business.'"

12. Mill, *Principles of Political Economy*, bk. 5, chap. 2, para. 28.

13. George, *Progress and Poverty*, bk. 5, chap. 2, para. 30.

14. George, bk. 8, chap. 2, para. 12–17.

15. Massey and Catalano, *Capital and Land*, 67–68, 114–38. They argued that this perspective was increasingly being adopted by industrial landowners as well. See also Whatmore, "Landownership Relations."

16. Harvey, *Limits to Capital*, 347.

17. Harvey, 368. For more on the treatment of land as a financial asset in relation to earlier theories of rent see Haila, "Theory of Land Rent," and Ward and Aalbers, "'Shitty Rent Business.'" For more on the treatment of land as a financial asset during the current land rush see Fairbairn, "'Like Gold with Yield'"; Gunnoe, "Political Economy of Institutional Landownership"; Knuth, "Global Finance and the Land Grab." Gunnoe ("Political Economy of Institutional Landowneship") describes growing finance-sector interest in acquiring farmland and timberland as evidence that we are living in a "neo-rentier society."

18. Harvey, *Limits to Capital*, 331, 369. Though Harvey admits that the positive functions of subjecting landownership to financial logic are "bought at the cost of permitting insane forms of land speculation" (331), he does not dwell on this point. In slightly sidelining speculation, he follows Marx, who also tended to exclude it from his analysis (367).

19. Graeber, "Value," 440–43.

20. The capitalization rate can be further disaggregated into its main components: the discount rate, which is basically the prevailing government interest rate adjusted upward to account for the greater risk of investing in farmland than in government bonds, and the rate at which farm income is expected to grow in coming years (Gloy et al., "Farmland Values," 11–13). See also J. Henderson, "Will Farmland Values Keep Booming?," 88–89 and note 4. For a really simple explanation and examples see S. Anderson, "Real Estate Valuation Strategy."

21. Low interest rates also indirectly affect farm income by reducing the exchange rate and therefore boosting the value of national exports.

22. The first equation also illuminates how land prices can impact agricultural production processes. When land prices rise relative to farm income, the capitalization rate will be relatively lower, making an investment in farmland less attractive. This means that high farm prices either put pressure on farmers to sell their land (since they could be making a higher rate of return by investing that money elsewhere) or to take steps to increase their farm income, intensifying their production process or cutting input costs such as wages paid to workers. It is thus not only true that high farm incomes lead to higher land values, but also that higher land values feed into the "treadmill of production," which pushes farmers to constantly try to increase their farm incomes (Guthman, "Back to the Land").

23. It is common to distinguish between booms and bubbles, but exactly when a boom becomes a bubble is a matter of some dispute. For most economists, *booms* occur when asset price increases are justified by the economic fundamentals, while *bubbles* occur when speculative impulses take over, inflating prices beyond what is justified by the fundamentals. Kindleberger and Aliber (*Manias, Panics, and Crashes*, 29) recommend the far narrower bubble definition of "an upward price movement over an extended period of fifteen to forty months that then implodes."

24. Wahl, "Food Speculation." Economists do, however, recognize that different forms of investment have differing orientations toward risk and time. A common distinction is made between three classes of investors: *hedgers*, who make investments in order to reduce a risk that they already face; *arbitragers*, who enter into simultaneous transactions in at least two different markets in order to make a riskless profit based on price discrepancies; and *speculators*, who make (at least somewhat) risky investments based on anticipated future price movements either up or down (Hull, *Options, Futures, and Other Derivatives*, 10–15).

25. Chryst, "Land Values and Agricultural Income," 1265; Turvey, "Hysteresis," 183. Some economists have been forced to conclude that, at least in the short term, land prices are affected by investment fads or speculative bubbles, a perspective in line with the growing subfield of behavioral economics (Falk and Lee, "Fads versus Fundamentals"; Roche, "Fads versus Fundamentals in Farmland Prices: Comment"; Shiller, "Understanding Recent Trends").

26. Schnitkey and Sherrick, "Income and Capitalization Rate Risk."

27. This discussion is indebted to Jean Tirole (not a post-Keynesian economist by any means), who stated that asset bubbles are most likely to occur when three conditions are present: "*Durability, scarcity,* and *common beliefs*" (Tirole, "Asset Bubbles and Overlapping Generations," 1521, original emphasis).

28. Featherstone and Moss, "Capital Markets, Land Values, and Boom-Bust Cycles"; Shiller, "Understanding Recent Trends."

29. The opposite is also true. Using rent/value ratios as an indicator, Lindert ("Long-Run Trends," 65, original emphasis) found "a *systemic pattern in forecast errors*: the further the real price of farm land is above (below) its previous longrun trend, the greater the likely overoptimism (overpessimism) about the subsequent price trend." The rent/value ratio is a common way of determining whether farmland markets are becoming

overheated. The logic is this: since rent acts as a proxy for farm income, if land prices were based purely on farm incomes, then rent and price would march pretty much in tandem, keeping the ratio steady over time. If land starts to become overvalued, then the ratio will fall; if land becomes undervalued, the ratio will rise.

30. For an excellent account of market booms and busts, which draws on the work of Minsky, see Kindleberger and Aliber, *Manias, Panics, and Crashes.*

31. Marx, *Capital,* vol. 1, preface, para. 8; Christophers, "For Real," 140, 145.

32. Rental income and production income are clearly not the same thing. The former is based on the latter, but there is a lag time before the rent charged adjusts to changes in commodity prices or other conditions. However, it was common among the people I interviewed to refer to both types of return as "income," "yield," or "cash returns."

33. McGrath, "Majority of Public Plans."

34. IRR is a common measure of the expected returns to be generated by a potential investment.

35. Farmland LP, Home page; Guthman, "Back to the Land."

36. Janiec, "Peoples Company."

37. This need not necessarily be the case. Sometimes subdivision can increase property value. One Brazilian fund manager told me that his company sometimes breaks up very large farms into 1,000- to 3,000-hectare chunks in order to increase the number of potential buyers.

38. Guthman, "Back to the Land," 525.

39. For a brilliant discussion of the concept of "highest and best use" in urban development see Blomley, "Mud for the Land."

40. As Jens Beckert (*Imagined Futures,* 133) puts it: "Investments are motivated by imaginaries of how the future will unfold. Actors express these imaginaries in the form of narratives that show their convictions, beliefs, fears, and hopes, supported by calculative tools."

41. Beckert, *Imagined Futures.* In a study of fund managers, Tuckett found that, faced with considerable uncertainty and imperfect information, they make decisions not by rationally weighing all information but by constructing shared narratives about likely investment outcomes. Their decisions about what to buy and sell are based less on the "fundamentals" of an investment than on the narratives they build around those fundamentals (Tuckett, "Financial Markets"). Likewise, Chong and Tuckett found that emotionally powerful "conviction narratives" are essential to overcoming uncertainty and inspiring investment. In fostering conviction about the future, investment narratives can alter the behavior of market actors, the value of commodities and companies, and, ultimately, the future (Chong and Tuckett, "Constructing Conviction").

42. Tsing, *Friction,* 57.

43. This point is made very eloquently by Tania Li. In her article "What Is Land?," she argues for viewing land as an assemblage of diverse elements, including such "inscription devices" as scarcity narratives and graphs of rapidly rising land prices. Oane Visser ("Running Out of Farmland?"), meanwhile, reveals that such narratives are often unsuccessful, leading to "value stagnation."

44. Austin, *How to Do Things with Words,* 4–7; Callon, "What Does It Mean to Say That Economics Is Performative?" and "Embeddedness of Economic Markets"; De Goede, *Virtue, Fortune, and Faith,* 5–8, 179–80; MacKenzie, *Engine, Not a Camera,* 15–25.

45. For more on scarcity narratives see Hartmann, "Ghosts of Malthus"; Hildyard, "'Scarcity' as Political Strategy."

46. Lappé, "Beyond the Scarcity Scare."

47. Kolesnikova, "Grantham Says."

48. Grantham, "Time to Wake Up."

49. McIntosh, "Aquila Capital."

50. Mehta, "Scare"; Ross, "Lonely Hour."

51. Sen, *Poverty and Famines*; Watts, *Silent Violence*.

52. Lappé, Collins, and Fowler, *Food First*, 22.

53. Unlike the other investment conferences mentioned in this book, I did not actually attend this particular event. However, a video of the speech is available in full online (Dotzour, "Peoples Company Land Expo Keynote").

54. Esposito, *Future of Futures*, 37.

55. Esposito, 13. According to Esposito, "The more one is free to construct one's own temporality, the more one must take the equal and yet opposite freedoms of others into account. The uncertainty of the future is multiplied by the uncertainty of the behavior of all other operators who are oriented to the same future" (Esposito, *Future of Futures*, 28).

56. Land Commodities, *Agriculture and Farmland Investment Report*, 131.

57. Colman and Daley, "Global Farmland Primer," 2.

58. Gustke, "Digging into Farmland."

59. Butler, "Performative Agency," 149.

60. Austin, *How to Do Things with Words*, 16; Butler, "Performative Agency"; Callon, "Performativity, Misfires and Politics."

61. An April 2019 article on the Agri Investor news site was dedicated to this issue (Janiec, "Is Ag Suffering"). Given that inflation hedging is one of the primary selling points for farmland, it reported, this absence of inflation is problematic for those raising capital for agricultural funds.

62. Visser, "Running Out of Farmland." See also Kuns, Visser, and Wastfelt, "Stock Market and the Steppe."

63. Fourcade, "Cents and Sensibility," 1722; Radin, *Contested Commodities*, xiii, 20–21.

64. Radin, *Contested Commodities*, 57–59, 102–14.

65. Anthropologists describe societies in which land and self are so tightly bound by the bones of buried ancestors and the sweat of ritual ceremonies that they are not fully separable (Povinelli, *Cunning of Recognition*). See also Verdery, *Vanishing Hectare*, and Shipton, *Mortgaging the Ancestors*, for connections between land, ancestry, kinship, and community identity.

66. Crucially, this does not mean that they resist commodification entirely; sacred things can and often do become commercialized without necessarily losing their sacred qualities (Zelizer, "Human Values and the Market").

67. Zelizer, "Human Values and the Market."

68. De Goede, "Discourses of Scientific Finance."

69. Mazzucato, *Value of Everything*, 102.

70. The intertwining of farmland value and moral values is discussed extensively in several recent articles, including Kish and Fairbairn, "Investing for Profit"; Ouma, "This Can('t) Be an Asset Class"; and Sippel, "Financialising Farming." Importantly, Sippel ("Financialising Farming," "Food Security or Commercial Business?") points out that the moral narratives used to legitimize farmland investment differ between national contexts. In Australia, which has long embraced a neoliberal approach to agriculture, it is not financial ownership of land, but foreign (and particularly Arab and Chinese) ownership that is open to moral challenge. In this context, Australian superannuation funds actually emerge as moral saviors of Australian farmland. Chapter 4 of this book considers a similar moral and political debate taking place in Brazil.

71. Because the concept of moral economy has most often been explored in relation to peasants, one might receive the impression that moral economies exist in opposition to free-market capitalism. They often become visible only when they are violated—say, by very steep hikes in land rent or grain prices—prompting outrage, protest, or even violence.

However, as Wendy Wolford ("Agrarian Moral Economies") has shown, powerful eco-nomic actors subscribe to their own moral economies, which may be perfectly compatible with neoliberal capitalism.

72. Hotel name altered to preserve confidentiality.

73. Gieryn, "Boundary-Work."

74. Larder, Sippel, and Lawrence, "Food Security Narratives," examines how agricul-tural investors position themselves as food producers in the Australian context so as to both attract capital and achieve moral legitimacy.

75. Locke, *Second Treatise of Government*, chap. 5, section 38.

76. Charles Geisler argues that "underutilization" is a crucial "new *terra nullius* narra-tive" used to validate the present wave of enclosures in the Global South (Geisler, "New Terra Nullius Narratives"). See also Baka ("Political Construction of Wasteland"), who describes how Locke's concept of "wasteland" has been deployed by the Indian govern-ment to justify large rural land acquisitions by the state.

77. Magnan, "Financialization of Agri-food," 10; Sommerville and Magnan, "'Pin-stripes on the Prairies,'" 128.

78. Bonnefield, "Four Sons and a Farm." Investors also often point to generational shifts in the farm population as a reason their services are needed—their farmland pur-chases, they argue, allow aging farmers to retire and to free up an inheritance for children who do not wish to farm (Sommerville and Magnan, "'Pinstripes on the Prairies,'" 128).

79. Heppner, "Reasons to Sell Land."

80. Sippel, Larder, and Lawrence, "Grounding the Financialization of Farmland," 258.

81. "When trade in land is reduced to a special branch of the circulation of interest-bearing capital, then, I shall argue, landownership has achieved its true capital-istic form. . . . Once such a condition becomes general, then all landholders get caught up in a general system of circulation of interest-bearing capital and ignore its impera-tives at their peril. Owner-producers, for example, are faced with a clear choice between purchasing the land or renting it from another. How that choice is exercised, under pure conditions of capitalist landownership, should make no difference" (Harvey, *Limits to Capital*, 347).

## 3. MATERIAL DIFFICULTIES

1. For more on efforts to "assetize" farmland and their (abundant) failures see Ducas-tel and Anseeuw, "Agriculture as an Asset Class"; Kuns, Visser, and Wastfelt, "Stock Market and the Steppe"; Li, "Rendering Land Investible"; Ouma, "Financialization to Operations of Capital" and "This Can('t) Be an Asset Class"; Visser, "Running Out of Farmland?"

2. Kautsky, *Agrarian Question*, 12 (original emphasis).

3. Another important part of the explanation is that peasant farmers are, in certain ways, able to outcompete capitalist producers. While capitalists must turn a profit or go out of business, peasant farmers have an ability to self-exploit, which allows them to better survive the periods of exceptionally low crop prices that often wrack the farm economy. They will work themselves (and their children) to the bone, going hungry if necessary, to avoid losing their farms. Ultimately, Kautsky believed, big capitalist producers would replace smallholder family farms, but it might take a long, long time (Kautsky, *Agrarian Question*).

4. Most famously, Mann and Dickinson argue that the disjuncture between produc-tion time and labor time in agriculture makes farming inherently unpalatable to capitalist producers. In temperate climates, the actual work of farming, during which value is cre-ated, can occur only at particular times of the year. The rest of the time is spent waiting for the winter to pass, waiting for plants to grow, or waiting for livestock to mature. This forced inactivity reduces the rate of profit, makes it harder to hire workers, and necessitates

considerable storage infrastructure if the company is to sell grain year-round. Mann and Dickinson also observe that, for those crops that cannot be stored, there are all the additional risks and time pressures associated with perishability (Mann and Dickinson, "Obstacles to the Development"). Agriculture presents a barnload of other issues as well, however. The physical extensiveness of agriculture makes it difficult to control a hired workforce and limits the potential for economies of scale. "New product development" is constrained by the pace and possibilities of plant and animal reproduction (or at least it was until biotechnology came along). Finally, farming involves major risks from weather, pests, and disease. For more on the various barriers to the capitalist penetration of agriculture see FitzSimmons, "New Industrial Agriculture"; Goodman, Sorj, and Wilkinson, *From Farming to Biotechnology*; Kloppenburg, *First the Seed*; Lewontin and Berlan, "Technology, Research, and the Penetration of Capital"; Mann, *Agrarian Capitalism*.

5. Boyd, Prudham, and Schurman ("Industrial Dynamics") argue that in all extractive and cultivation-based industries, nature presents industrial capital with obstacles but also opportunities and surprises.

6. Goodman, Sorj, and Wilkinson, *From Farming to Biotechnology*.

7. Mooney, "Labor Time, Production Time," 280, 289.

8. G. Henderson, *California and the Fictions of Capital*, 29.

9. To see the financial sector making hay out of agricultural risk, one need look no further than crop insurance and derivative markets, both of which were created to protect farmers from risk but subsequently became profitable financial realms in their own right. Agricultural derivatives reduce the uncertainty farmers and food processors experience in the long months between planting and harvesting by letting them agree in advance on the price to be paid for the crop regardless of what might happen in the intervening period (Clapp, *Food*; Clapp and Helleiner, "Troubled Futures?"; Cronon, *Nature's Metropolis*). Crop insurance deals directly with weather risk by providing financial protection against the damage that can result from weather events such as drought, hail, frost, and excessive rain. Now a growing market in "weather derivatives" has further commodified weather risk by abstracting it from specific geographic locations and damages. Whereas traditional weather insurance requires the holder to demonstrate an insurable interest (some kind of property or business that would be affected by the weather in question) and proof of loss (evidence that the damage was caused by the weather event), the price of weather derivatives is instead based on underlying meteorological indices such as temperature and rainfall, allowing for hedging and speculation on even mundane variations in the weather (Pollard et al., "Firm Finances"; Randalls, "Weather Profits"). Along similar lines, development institutions are increasingly encouraging small farmers to buy "index insurance," an agricultural micro-insurance product that makes payments based on agricultural and weather-related data, making it more like a weather derivative than a traditional insurance product (Breger Bush, *Derivatives and Development*; L. Johnson, "Index Insurance"; Isakson, "Derivatives for Development?").

10. G. Henderson, *California and the Fictions of Capital*, x.

11. Economists differentiate between risk (when future outcomes are unknown but their likelihood can be estimated) and uncertainty (when too much is unknown to evaluate the odds) (Knight, *Risk, Uncertainty and Profit*); I am not particularly consistent in maintaining this distinction. Importantly, as suggested in chapter 2, economic uncertainty is not necessarily a bad thing; there is always the possibility that a favorable future will bring windfall profits; the word *windfall* itself suggests an upside to natural uncertainty.

12. Goodman, Sorj, and Wilkinson, *From Farming to Biotechnology*, 156.

13. Friedland, Barton, and Thomas, *Manufacturing Green Gold*, 44.

14. The risks associated with landownership and agricultural production cannot be neatly categorized as either "natural" or "social." The two are interlinked as, for instance,

when a drought leads to high crop prices, causing farmers to respond en masse by increasing production and contributing to a period of oversupply. That "natural" phenomena like droughts are intensified by anthropogenic climate change adds a layer of interconnection. (On the social-ness of all "natural" disasters see, for example, Klinenberg, "Denaturalizing Disaster"; Bakker, "Katrina"; Braun and McCarthy, "Hurricane Katrina and Abandoned Being.")

15. Ducastel and Anseeuw, "Agriculture as an Asset Class," 204; Ouma, "Financialization to Operations of Capital," 85.

16. MacKenzie, *Engine, Not a Camera*, 45–54.

17. As Ouma ("Financialization to Operations of Capital," 85) points out, MPT has yet to be applied to farmland in the highly mathematical ways that it is with other asset classes. This, he argues, has three causes: lack of historical data on farmland prices in many locations, lack of daily price data such as that available for stocks, and the fact that "many farmland investments are conducted on a case-to-case basis rather than through sophisticated portfolio structures" (Ouma, "Financialization to Operations of Capital," note 8).

18. Kaplan, "Farmland as a Portfolio Investment," 4.

19. Emphasis added.

20. Gliessman and Engles, *Agroecology*, 203–11; Mann, *Agrarian Capitalism*, 64. Large growers in places like California have also long used crop diversification to reduce market risk and increase their appeal as retail suppliers (Walker, *Conquest of Bread*, 93).

21. Multilateral Investment Guarantee Agency, "MIGA and Chayton Capital" and "MIGA Reinsures Agribusiness."

22. Espeland and Stevens, "Commensuration as a Social Process." Commodification depends on commensuration, on a very basic level, because it requires that disparate types of value all be translated into price. When people wish to resist commodification, meanwhile, they often do so on grounds of incommensurability; they argue that certain types of moral and cultural value cannot be captured by a cost-benefit analysis or that the thing in question is simply "priceless." Espeland ("Value-Matters") and Schmelzkopf ("Incommensurability") offer detailed accounts of disputes over the commensurability of land.

23. Savills Research, "Global Farmland Index."

24. Savills Research, "International Farmland Focus 2012." In this publication the tool is called the Risk-Return Matrix.

25. Stark ("For a Sociology of Worth," 5) makes this point beautifully: "I propose that we develop a concept of accounts. Etymologically rich, the term simultaneously connotes bookkeeping and narration. Both dimensions entail evaluative judgments, and each implies the other: Accountants prepare story lines according to established formulae, and in the accounting of a good storyteller we know what counts. In everyday life, we are all bookkeepers and storytellers. We keep accounts and we give accounts, and most importantly, we can be called to account for our actions."

26. Fritz, "Illinois Pension Board"; 2018 interview with NCREIF official.

27. While helpful for raising investor capital, these optimistic projections may cause problems for the agricultural fund managers who use NCREIF to benchmark their own performance. At a 2015 meeting, for instance, board members of the Orange County Employees Retirement System (OCERS) demanded to know why their US farmland investments were "grossly underperforming" relative to the NCREIF index. Representatives of two different farmland investment companies were forced to explain that NCREIF was not a good benchmark for their funds, one adding that they therefore benchmark their performance to a more modest, "bespoke" benchmark they create from the NCREIF data to more closely resemble their portfolio. Investor-manager negotiation over benchmarks and its significance for farmland's construction as a financial asset class is thoughtfully explored by Ducastel and Anseeuw ("Agriculture as an Asset Class").

28. Hemphill, "Australian Farmland Index."

29. See Li, "Centering Labor in the Land Grab Debate" and "What Is Land?," for an insightful and far more in-depth critique of this report.

30. Goldman, *Imperial Nature*.

31. Wolf and Wood, "Precision Farming."

32. Carolan, "Publicising Food"; Bronson and Knezevic, "Big Data in Food and Agriculture."

33. Rogers, "How Agtech Has Quietly Transformed."

34. However, scholars have pointed out that, under the right social conditions, digital data could also facilitate agroecological production, small farmer cooperatives, and other outcomes that subvert the status quo of agricultural consolidation (Carolan, "Publicising Food"; Rotz et al., "Politics of Digital Agricultural Technologies").

35. CiBO Technologies, "About."

36. Tillable, "For Farmland Investors."

37. There is a very large literature on big data and surveillance, but see, for example, Andrejevic, "Surveillance in the Digital Enclosure," and Zuboff, "Big Other." On the concept of "data subjects" see L. Taylor, "Data Subjects or Data Citizens?"

38. Kautsky, *Agrarian Question*, 147.

39. Payment of the carried interest is often made contingent on the fund surpassing a certain predetermined internal rate of return called the "preferred return" or "hurdle rate." Once the fund has been exited, a "distribution waterfall" determines the order in which everyone gets paid, roughly as follows: first the investors (limited partners) are paid back up to this preferred return rate, then the asset manager (general partner) is paid back up to the preferred return, then any remaining profits are split, with most going to the investors and the remainder going to the asset manager in the form of carried interest.

40. For example, Swiss fund of funds Adveq bought 18,000 ha of Australian almond orchards—fully half of Australia's almond land—through a co-investment with the Municipal Employees' Retirement System of Michigan and Danish fund Danica Pension (Dunkley, "Investors Are Going Nuts for Nuts").

41. Kuns, Visser, and Wastfelt ("Stock Market and the Steppe," 204) describe assembling small landholdings as an "administrative hassle" for the large agricultural operating companies working in Russia and Ukraine.

42. For a fascinating discussion of the obstacles faced by publicly traded agricultural operating companies in Ukraine and Russia see Kuns, Visser, and Wastfelt, "Stock Market and the Steppe."

43. Fligstein, "End of (Shareholder Value)"; Lazonick and O'Sullivan, "Maximizing Shareholder Value."

44. Gunnoe and Gellert, "Financialization."

45. SLC Agrícola, "December 2012 Presentation for Investors."

46. In 2016, Cosan sold its shares in Radar to TIAA. The web page from which this quote is drawn is therefore no longer available. It was originally cited in Fairbairn, "Like Gold with Yield."

47. Burch and Lawrence, "Financialization in Agri-food"; Christophers, "On Voodoo Economics." Christophers, drawing on Marx and Harvey, argues that this move rests on a "mystification"; it asserts a false divide between the property and the activities that property is able to support, when in fact the two are innately entangled.

48. Krippner, *Capitalizing on Crisis*, 28–29.

49. This is a description of an equity REIT. There are also mortgage REITs, which invest in mortgages and mortgage derivatives and disburse dividends based on these debt repayments.

50. The Bulgarian REIT equivalents, known as Special Purpose Investment Companies, were made possible with the passage of a 2003 act that exempted these entities from corporate tax provided they, like US REITs, distribute 90 percent of income to investors (Stooker, *REITs around the World*, 81). REITs were established in the US with the passage of the Real Estate Investment Trust Act of 1960 (Han and Liang, "Historical Performance").

51. Gladstone Land, *2016 Annual Report*.

52. Farmland Partners, *2016 Annual Report*.

53. Rural Funds Group, "RFF Fund Overview" and "Assets."

54. Cortese, "Crowdfunding Crowd."

55. July 2013 interview with Fquare executive.

56. AcreTrader, "How It Works."

57. Christophers, "For Real," 145.

58. Marx, *Theories of Surplus Value*, pt. 3, addendum.

59. Ouma ("Getting in between M and M'") cautions strongly against overstating this point. He argues that scholars should focus on unpacking the practices and frictions that occur in between M and M', which he rightly notes are too often black-boxed in financialization scholarship.

60. This same speaker also mentioned that his fund had hired a sunspot expert to help them predict the future weather and its impact on agriculture. This expert, he said, had successfully used sunspots to predict a drought in Russia and a period of increased rain in Australia for his fund. But the sunspot expert wasn't just a weatherman. The hedge-fund manager next told the audience that his expert had predicted a major solar flare for the coming week—and a resulting stock-market downturn, due to how the sun's magnetic energy affects human emotions. He offered this stock-market tip in a playful spirit, but the first audience member to comment after the presentation said that his company also had a "sunspot guy," and the two had a back-and-forth about the implications of sunspots and solar flares for market movements. This interchange was striking given the somewhat ignominious history of sunspots in economic thought. William Stanley Jevons, the British economist whose work on marginal utility theory helped launch the neoclassical transformation of economics, published an article in *Nature* in 1878 titled "Commercial Crises and Sun-Spots," in which he attempted to explain the business cycle via solar activity. This argument was not substantiated by subsequent research and did little for Jevons's reputation as a great economic mind. Beginning in the mid-1980s, however, sunspots have had a second career as a financial metaphor. An influential 1983 paper by David Cass and Karl Shell—making tongue-in-cheek reference to Jevons's discredited theory—used the term as a stand-in for "extrinsic uncertainty," or random variables that do not affect market fundamentals but may nonetheless cause shifts in investor sentiment and therefore shifts in markets. Their article—which was titled "Do Sunspots Matter?"—found that changes to investor expectations could cause price movements, even though the fundamentals of the economy stayed the same. In other words, sunspots matter if people believe that they do (Cass and Shell, "Do Sunspots Matter?"). Given that sunspots (the literal kind) have been essentially debunked as an economic variable, it is telling that anyone would still seek the advice of a sunspot consultant. It is, first, a testament to the enduring uncertainty of agriculture—no amount of diversification or hedging can take the place of favorable weather in ensuring that agriculture-based ventures are profitable. Second, the suggestion, however skeptically made, that sunspots can be used to predict market movements is a signal of the additional uncertainty introduced by finance. Though agriculture has always been subject to natural unpredictability, the increasing financialization of agriculture introduces a new source of uncertainty in the form of financial markets themselves.

61. Gustke, "Farm to Market."

62. The stock prices of the three US farmland REITs may be particularly volatile because of their size. All three have very low market capitalizations, meaning that a sale of just a thousand shares can cause their value to fall (Gustke, "Farm to Market").

63. Rota Fortunae, "Farmland Partners."

64. De Goede, *Virtue, Fortune, and Faith*, 29; Rescher, *Luck*, 11.

65. Short selling works like this: the short seller borrows stock, sells it at its current price, and then, once the stock price has fallen, buys it back at the lower price and returns it to its owner. By selling high and then buying low, the short seller makes a profit.

66. Farmland Partners, "Farmland Partners Files Lawsuit."

67. Ulrich Beck argues that modern society, through its technological advances and rational attempts to control the natural world, has paradoxically created new and far more insidious forms of risk; climate change, the spread of radiation after a nuclear melt-down, and the growth of antibiotic-resistant "superbugs" are just a few examples (Beck, *Risk Society*). It has likewise been argued that today's financial crises are a product of the increasingly complex and interconnected institutions of modern finance, including the derivatives that were initially created as a means to reduce market risk (LiPuma and Lee, *Financial Derivatives*).

68. Kuns, Visser, and Wastfelt ("Stock Market and the Steppe") found that the poor performance of publicly traded "agroholding" companies in Russia and the Ukraine stemmed in part from their managers sacrificing long-term agricultural development in an attempt to meet stock market demand for short-term profits.

## 4. FOREIGN POLITICS

1. In this case, because the company is publicly held, the financial institutions that sent representatives to tour the farm were likely engaged in trading or analyzing the company's securities.

2. Data from Censo Agropecuário 2006 cited in Sauer and Leite, "Agrarian Structure," 877.

3. Osório Silva, *Terras devolutas e latifúndio*; Panini, *Reforma agrária*.

4. Imperial Law 601 (September 18, 1850).

5. Osório Silva, *Terras devolutas*. This was not entirely out of the blue. In the 1690s the Portuguese began to charge a property tax on *sesmarias*, contributing to a gradual reconceptualization of them as private-property rights carved from the public domain, rather than just usufruct rights (Lima, *Pequena história*). But it was only in 1850 that a land market was officially created. Among other things this meant that, for the first time, land could now be used as collateral on loans (Panini, *Reforma agrária*).

6. Osório Silva, *Terras devolutas*.

7. Brannstrom, "Producing Possession."

8. Dias, Vieira, and Amaral, *Comportamento do mercado*.

9. Reis, "Brazil: One Hundred Years."

10. Reis, "Brazil: One Hundred Years."

11. Reis, "Brazil: One Hundred Years"; Skidmore, *Brazil*.

12. A. Hall, "Land Tenure and Land Reform"; Welch, "Globalization and the Transformation of Work." The Land Statute was Law 4.504 (November 30, 1964).

13. De Sousa and Busch, "Networks and Agricultural Development."

14. Graziano da Silva, *A modernização dolorosa*; Delgado, *A questão agrária*.

15. Ankersen and Ruppert, "Tierra y Libertad"; Cunha, "Social Function of Property"; Pereira, "A teoria da função social."

16. Wolford, *This Land Is Ours Now*, 37–38.

17. Pereira, "A teoria da função social"; Wittman, "Reframing Agrarian Citizenship."

18. A. Hall, "Land Tenure and Land Reform."

19. Welch, "Globalization and the Transformation of Work"; Wittman, "Reframing Agrarian Citizenship."

20. Carter, "Broken Promise."

21. Reydon, "A regulação institucional"; Reydon and Plata, *Intervenção estatal*; Reydon, Fernandes, and Telles, "Land Tenure in Brazil."

22. Dias, Vieira, and Amaral, *Comportamento do mercado*; Reydon, "A regulação institucional."

23. The regime offered subsidized rural credit, which, given high rates of inflation, often carried a negative real interest rate; farmland ownership was one of the primary ways to tap into this stream of easy money, increasing its appeal and price (Delgado, *Capital financeiro e agricultura*; Rezende, "Crédito rural subsidiado"). At the same time, the government's policy of extending infrastructure into frontier regions created opportunities for real estate speculation and increased the price of land already accumulated by elites. The military regime also offered considerable tax breaks to those who undertook cattle ranching in the Amazon region, which contributed to pastureland valorization even in the face of declining soil productivity (Hecht, "Environment, Development and Politics"). In addition to all these policy inducements to land speculation, land prices were benefited by a Brazilian stock-market crash in 1971, which sent the wealthy flocking back to land as a trusted reserve of value (Sayad, "Preço da terra").

24. Land prices leapt in 1986 when the government of José Sarney decided to tackle inflation by replacing the existing currency, the cruzeiro, with a new unit known as the cruzado. *Plano Cruzado*, as it was called, froze all prices and wages at their current levels, leading to a considerable cooling of financial markets. In search of alternative investments, capital poured into farmland markets, causing prices to spike, though they quickly dropped again as *Plano Cruzado* was abandoned and investment in financial markets returned. Land prices also shot up in 1989 when investors became anxious about the possibility that a labor leader—future president Lula—might win the presidency. Then that same year an insider-trading scandal almost crashed the Rio de Janeiro stock market (the Nagi Nahas scandal), deepening the crisis of confidence in financial markets. Another land price spike occurred in 1992 when financial instability caused by hyperinflation was joined by political instability as President Fernando Collor de Mello resigned just hours before the Senate was set to impeach him on corruption charges (Reydon, Anãnã, et al., "Ativo terra agrícola," 187–93).

25. Wilkinson, Reydon, and Di Sabbato, "Dinâmica no mercado de terras."

26. Notes on figure 4.3: Nominal prices converted to 2014 dollars using the Índice Geral de Preços—Disponibilidade Interna (IGP-DI).

27. Sauer and Leite, "Agrarian Structure"; Wilkinson, Reydon, and Di Sabbato, "Concentration and Foreign Ownership."

28. McMichael, "Land Grabbing as Security Mercantilism," 48.

29. Trevisan, "Estatal da China"; Sant'Anna, "Chineses desistem." See G. Oliveira, "Chinese Land Grabs in Brazil?" for a detailed discussion of the various reported Chinese agricultural investments in Brazil.

30. Agrolink, "Discreta, tiba agro investe pesado."

31. Zhou, "Brookfield Nears First Close."

32. Nuveen, *Responsible Investment in Farmland*, 2018.

33. Adecoagro, "Institutional Presentation"; Adecoagro, "Company/Organizational Structure."

34. BrasilAgro, "Institutional Presentation."

35. *Territory* refers to land's frequent association with group identity (whether religious, ethnic, or national) and the attendant idea that the leaders of that group should exercise some kind of authority over the area in question (D. Hall, *Land*). The concept of the "nation-state," for instance, expresses this connection between a cultural or ethnic group (the nation) and a political entity (the state) via association with a particular geographic territory (the land). Though sovereign state control over national territory has been challenged somewhat by globalization (Hudson, "Beyond the Borders"; Paasi, "Boundaries as Social Processes"), the idea of national territory and borders remains very powerful.

36. Cardoso and Faletto, *Dependency and Development*; Evans, *Dependent Development*.

37. A. Oliveira, "A questão da aquisição."

38. Garrido Filha, *O Projeto Jari*, 87. Some of these land acquisitions enjoyed the protection of the law and were even smiled upon by the government; at the invitation of the military leader, an American billionaire named Daniel Ludwig acquired 5 million hectares—an area roughly the size of Costa Rica—in the Brazilian state of Pará for a cellulose project called Projeto Jari (Garrido Filha, *O Projeto Jari*; Sautchuk, Martins de Carvalho, and Buarque de Gusmão, *Projeto Jari*). Other acquisitions were blatantly illegal, such as the millions of hectares acquired by an American named Stanley Amos Selig through an enormous operation of *grilagem* (Lindoso, "CPI comprova irregularidades"). Among other exploits, the Velloso Report states that Selig succeeded in acquiring an entire municipality—Ponte Alta do Norte—in the state of Goiás. This municipality was 1,305,000 ha in size, but Selig oversold the land titles, with lots sold to Americans totaling 1,390,438 ha (Garrido Filha, *O Projeto Jari*, 85). It was reported in 1970 that Selig had been murdered by a defrauded American investor (*Estado de São Paulo*, "Mataram Selig").

39. An initial attempt at regulating foreign land purchases—Supplementary Act No. 45, created in January 1969—restricted the acquisition of rural land to Brazilians and foreigners who were residents of Brazil. Law 5.709 (October 7, 1971) was regulated by Decree 74.965 (September 26, 1974).

40. The key section of the law is article 1, section 1 of Decree 74.965, which states, "Also subject to the regime established by this regulation is a Brazilian company in which participate, in any capacity, foreign persons or companies that hold the majority of its capital and reside or are headquartered abroad." Although the restrictions initially only applied to land purchases, Law 8.629, passed in 1993, extended them to land leasing as well (Hague, Peixoto, and Filho, *Aquisição de terras por estrangeiros*, 9). Additionally, Law 6.634, passed in 1979, created a 150 km border zone in which the use of land was even more strictly controlled. In this area, foreign individuals and companies were prohibited from acquiring land, unless prior approval had been granted by the National Defense Council, then called the Council on National Security. In order to be considered Brazilian by this law, a company had to have 51 percent or greater Brazilian capital and at least two-thirds Brazilian employees (Wilkinson, Reydon, and Di Sabbato, "Dinâmica no mercado de terras").

41. Brown and Purcell, "There's Nothing Inherent about Scale"; Evans, *Dependent Development*.

42. Amann and Baer, "Neoliberalism and Its Consequences"; Mollo and Saad-Filho, "Neoliberal Economic Policies."

43. Câmara dos Deputados, "Relatório da Subcomissão." An initial, abortive attempt to repeal the law (attorney general Opinion GQ-22) argued that under the new constitution of 1988, it was no longer constitutional to discriminate against foreign-owned Brazilian companies on the basis of the source of their capital. This opinion was never

published in the Official Federal Gazette (*Diário Oficial da União*) and therefore did not go into effect (Hague, Peixoto, and Filho, *Aquisição de terras por estrangeiros*, 12–14). The 1998 attorney general opinion (Opinion GQ-181), which did go into effect, was based on Constitutional Amendment No. 6, passed in 1995, which had repealed an article of the 1988 Brazilian Constitution allowing for the privileging of "Brazilian companies of domestic capital" over non-Brazilian companies and foreign-owned Brazilian companies (Hague, Peixoto, and Vieira Filho, *Aquisição de terras por estrangeiros*, 15–17). The privileging of domestic enterprise had been a cornerstone of Brazil's import substitution industrialization approach to development, and so this amendment had widespread effects, allowing foreign capital to enter such formerly protected sectors as public utilities and oil exploration (Amann and Baer, "Neoliberalism and Its Consequences").

44. Between 2004 and 2005, Stora Enso purchased over 45,000 ha in the southern Brazilian state of Rio Grande do Sul, close to the borders with Uruguay and Argentina (Vieira, "Stora Enso obtém aval"). It easily complied with the toothless iteration of Law 5.709 then in effect by purchasing the land through a wholly owned Brazilian subsidiary, Derflin Agropecuária. This was a perfectly legal, and indeed normal, way to get around the restrictions on foreign landownership. However, because of the property location, Stora Enso found itself in violation of a second law (Law 6.634) requiring foreigners to get consent from the National Defense Council before buying land within 150 km of Brazil's terrestrial borders. To circumvent this obstacle, Stora Enso got even more creative. It established yet another Brazilian company, Azenglever Agropecuária, this time registered in the name of two of the company's Brazilian executives. Azenglever's land purchases were funded by a multimillion-dollar "loan" from Stora Enso, for which the land itself served as collateral (Vaz, "Aproveitando a flexibilidade"). Though this case largely catalyzed the 2010 reregulation, there was a happy ending for Stora Enso. Ultimately, the company was able to bypass INCRA and negotiate directly with the National Defense Council, which agreed to grant a retroactive "prior authorization" for the land purchase (Lerrer and Wilkinson, "Impact of Restrictive Legislation").

45. Quoted in Ogliari, "Compra de terras por múlti," A10.

46. Brazilian Senate, "Ata da 4a reunião," 368.

47. Stora Enso, "Shareholders and Ownership Changes."

48. Scolese, "Por dia, estrangeiro compra '6 Mônacos'"; Odilla, "Estrangeiros compram 22 campos."

49. Alvim, "Investimentos estrangeiros"; Hernandes, "Estudo sobre processos"; Pretto, "Imóveis rurais." In analyzing its own data, INCRA found that its national database contained a relatively meager 34,632 foreign-owned properties, totaling 4,037,667 ha (Pretto, "Imóveis rurais"). This land was concentrated primarily in the *cerrado* and was mostly registered to nationalities with a history of migration to Brazil, such as Portuguese, Japanese, and Italians (Wilkinson, Reydon, and Di Sabbato, "Dinâmica no mercado de terras").

50. Although it was prepared in 2008, this new attorney general's opinion (CGU/AGU Opinion No. 1/2008-RVJ) was not actually published (and therefore did not go into effect) until August 2010.

51. *Brasil de Fato*, "Assentamentos"; Peres and Dilorenzo, "Brazil's Indigenous Protest"; Rochedo et al., "Threat of Political Bargaining."

52. The effort to repeal the restrictions has been attempted under multiple bills, including Projeto de Lei (PL) 4.059/12 and, most recently, PL 2.963/19. See Zaia, "Casa Civil quer venda de terra"; Caetano, "Porteira aberta para os estrangeiros"; Zaia and Exman, "Ruralistas tentam emplacar."

53. *Valor Econômico*, "MST promete invasões."

54. The reasons given are as follows:

> "a) expansion of the agricultural frontier with the advance of cultivation in environmental protection areas and conservation units;
>
> b) irrational land price appreciation and incidents of real estate speculation generating an increase in the cost of the expropriation process for agrarian reform, as well as a reduction in the stock of land available for this end;
>
> c) increase in the illegal sale of public lands;
>
> d) use of funds from money laundering, drug trafficking and prostitution in the acquisition of land;
>
> e) increase in land grabbing [*grilagem*];
>
> f) proliferation of "front men" [*laranjas*] in the acquisition of these lands;
>
> g) increase in biopiracy in the Amazon Region;
>
> h) enlargement, without due regulation, of the production of ethanol and biodiesel;
>
> i) acquisition of land in the frontier zone putting national security at risk."

55. Alston, Libecap, and Mueller, *Titles, Conflict, and Land Use*, 51–2. In the Brazilian Constitution, the social function of land is defined as follows:

> "Article 186. The social function is met when the rural property complies simultaneously with, according to the criteria and standards prescribed by law, the following requirements:
>
> 1. rational and adequate use;
> 2. adequate use of available natural resources and preservation of the environment;
> 3. compliance with the provisions that regulate labour relations;
> 4. exploitation that favours the well-being of the owners and labourers."

56. Sauer and Leite, "Agrarian Structure."

57. While most of the people I interviewed are identified by pseudonyms, in the case of publicly elected politicians, I asked permission to use their real names.

58. De Sartre and Taravella, "National Sovereignty vs. Sustainable Development."

59. Wilkinson, Reydon, and Di Sabbato, "Dinâmica no mercado de terras." This position was reiterated to me in a June 2012 interview with a CNA official. Likewise, according to an agribusiness consultant I interviewed in November 2011, Glauber Silveira, the head of the Brazilian Association of Soy Producers (Aprosoja) had initially been in favor of the restrictions until this consultant and others convinced him otherwise.

60. In the end, the *ruralista*-dominated subcommittee voted overwhelmingly in favor of the Montes proposal. However, the resulting bill, PL 4.059/12, has not gone anywhere (see note 52).

61. G. Oliveira, "Chinese Land Grabs in Brazil?"; Brautigan, *Will Africa Feed China?* Likewise, in Australia, farmland investments by Arab investors (as well as Chinese investors) have received disproportionate scrutiny and been exposed to more public backlash (Larder, Sippel, and Lawrence, "Food Security Narratives"; Sippel, "Financialising Farming").

62. Said, *Orientalism*.

63. Another possible reason for the focus on Chinese investors could relate to the way that Chinese entities have gone about looking for land in Brazil. A Brazil-based farmland fund manager described them to me as "like an elephant in a crystal shop." Similarly, an agribusiness consultant, who was the only person I met who had firsthand experience of working with Chinese investors, told me that they "make a lot of noise in the market."

Rather than just hiring the best Brazilian consultants and trusting them to buy the land, he explained, they create a stir in their exhaustive consultation and search process. See G. Oliveira, "Chinese Land Grabs in Brazil?" for a thorough analysis.

64. Sant'Anna, "Chineses desistem."

65. MB Agro and Agroconsult, "Impactos econômicos do parecer"; Abraf, "Relatório de impactos negativos."

66. Agrifirma Brazil, "$82m Investment by Private Equity."

67. Interview with SLC Agrícola executive.

68. SLC Agrícola, "December 2012 Presentation for Investors."

69. This organizational structure was originally reported by researchers from Brazil's Social Network for Justice and Human Rights and GRAIN (Rede Social de Justiça e Direitos Humanos et al., "Foreign Pension Funds and Land Grabbing"), and the story was later picked up by the *New York Times* (Romero, "TIAA-CREF, US Investment Giant"). For more on the role of financial capital in Brazilian farmland markets and agriculture see Mendonça and Pitta, "International Financial Capital."

70. TIAA-CREF, *Responsible Investment in Farmland*, 2012; Nuveen, *Responsible Investment in Farmland*, 2018.

71. Holston, "Misrule of Law," 695. Holston's interpretation was certainly borne out in the case of Stora Enso, which despite flouting legal restrictions on foreign land acquisitions near the border, ultimately had its claim retroactively legalized.

72. Reydon, Introduction, 15–16. Translation by author.

73. Sauer and Leite, "Agrarian Structure." Even government policies that ostensibly fortify smallholder land rights—such as the *Terra Legal* program of land regularization in the Amazon—may simultaneously encourage land concentration by creating a more legible landscape for agribusiness expansion (G. Oliveira, "Land Regularization in Brazil").

74. A. Oliveira, "A questão da aquisição."

75. Brandão, "A formação territorial," 42–43.

76. Ofstehage, "Farming Is Easy."

77. Prager and Milhorance, "U.S. Investment Spurs Land Theft"; GRAIN and Rede Social de Justiça e Direitos Humanos, "Harvard's Billion-Dollar Farmland Fiasco"; McDonald and Freitas, "Harvard's Foreign Farmland Investment Mess."

78. Rede Social de Justiça e Direitos Humanos et al., "Foreign Pension Funds and Land Grabbing"; Romero, "TIAA-CREF, U.S. Investment Giant."

79. As Jennifer Clapp ("Financialization, Distance") puts it, financialization causes "distancing" within the global agri-food system, which makes it difficult to link causes with effects.

80. Hodgson, Cullinan, and Campbell, *Land Ownership and Foreigners*. Additionally, even in instances with no direct regulation of foreign ownership, indirect control may be exercised through differential tax treatment or other means.

81. Hodgson, Cullinan, and Campbell, *Land Ownership and Foreigners*; Tan, "Restrictions on the Foreign Ownership."

82. Morrison and Krause, *State and Federal Legal Regulation*. How a "foreigner" is defined is an important point in all these laws. When it comes to individual foreign investors, the criteria on which restrictions are based may be citizenship or residency (i.e., in some places, "aliens" are permitted to own land if they are permanent residents). When it comes to foreign corporations, the rules are much more complicated but are often based on incorporation outside of the US or percentage foreign ownership. The US also has federal laws in place that require reporting of foreign land acquisitions (DeBraal and Krause, "Corporate, Foreign, and Financial Investors"). Under pressure from the farm lobby, Congress passed the International Investment Survey Act (IISA) in

1976, followed by the Agricultural Foreign Investment Disclosure Act (AFIDA) in 1978 (Mason, "'PSSST, Hey Buddy'"). Neither of these laws prohibits foreign landowner-ship, but they do contain stringent reporting requirements. Under AFIDA, all foreign farmland holdings must be reported to the secretary of agriculture, who must provide regular reports on the extent of these holdings to the president, Congress, and the states (DeBraal and Krause, "Corporate, Foreign, and Financial Investors"). In 1980 Congress also passed the Foreign Investment in Real Property Tax Act (FIRPTA), closing a tax loophole that had allowed foreigners to buy and sell farmland without paying capital gains taxes (Mason, "'PSSST, Hey Buddy'").

83. GRAIN, "Land Ceilings."

84. In the US, for instance, antiforeign and anticorporate laws have often arisen in tandem and in response to the same land booms and busts. As Wilson ("Reforming Alien Agricultural Landownership," 713) puts it, "The two strands of agricultural law make for easy bedfellows—both ultimately based on a genuine desire to protect family farms from 'outsiders.'"

85. Wolford, "Agrarian Moral Economies."

## CONCLUSION

1. Average farm size in Malawi is 2.25 ha overall, and even less for women farmers (Fisher and Kandiwa, "Can Agricultural Input Subsidies," 103).

2. Windsor, "Corporate Social Responsibility." This perspective was forcefully cap-tured by Milton Friedman, who repudiated the idea of corporate social responsibility as "pure and unadulterated socialism," asserting that "the [only] social responsibility of busi-ness is to increase its profits" (Friedman, "Friedman Doctrine," 32–33).

3. US Congress, "Investment of Pension Funds in Farmland," 4. This hearing was called to consider the American Agricultural Investment Management Company (AAIMC), which closely followed on the heels of Ag-Land Trust. See chapter 1, note 60.

4. US Congress, "Investment of Pension Funds in Farmland," 5.

5. Gage, "Farmland Partners Helps Stabilize."

6. Sippel ("Financialising Farming," 549) found that in Australia, where political and moral concerns about farmland investment center on foreign government involvement, domestic superannuation funds were seen as fulfilling a "moral imperative" to keep land in "national hands."

7. Long, "Nearly Half of America."

8. G. Davis, *Managed by Markets*, 6, original emphasis.

9. G. Davis, 242.

10. Carolan, "Barriers to the Adoption."

11. Gliessman and Engles, *Agroecology*, 11–12.

12. Nair et al., "Carbon Sequestration in Agroforestry Systems."

13. Guthman, "Back to the Land."

14. FAO et al., "Principles for Responsible Agricultural Investment."

15. De Schutter, "How Not to Think about Land-Grabbing"; Stephens, "Principles of Responsible Agricultural Investment."

16. Committee on Food Security, *Voluntary Guidelines on the Responsible Governance of Tenure*.

17. See Clapp, "Responsibility to the Rescue?"; GRAIN, "Responsible Farmland Invest-ing?" and "Socially Responsible Farmland Investment," for more comprehensive lists and analysis of voluntary initiatives.

18. Principles for Responsible Investment, *Responsible Investment in Farmland*.

19. GRAIN, "Responsible Farmland Investing?" and "Socially Responsible Farmland Investment."

20. Clapp, "Responsibility to the Rescue?," 229–33. See also Clapp, "Financialization, Distance."

21. Borras and Franco, "From Threat to Opportunity?," 519.

22. Borras and Franco, 520.

23. For a discussion and critique of the CFS rai principle development process see Kay, "Political Brief."

24. De Schutter, "How Not to Think about Land-Grabbing."

25. Other coalition members included Family Farm Defenders, Wisconsin Farmers Union, Organic Consumers Association, Fair World Project, Rural Coalition, Presbyterian Hunger Program-PC (USA), MaryKnoll, Grassroots International, Faculty/Staff Divestment Network, 350NYC, Union of Concerned Scientists, Fian International, Fian Sweden, Other Worlds, Sum of Us, Food and Water Watch, and Responsible Endowments Coalition.

26. Quotes come from video of the rally obtained on the FOE US Facebook page.

27. Interview with campaign organizer; Grassroots International, "Tell TIAA."

28. Nuveen, *Responsible Investment in Farmland*, 2017, 2.

29. J. E. Davis, *Origins and Evolution*; Gray, "Community Land Trusts."

30. J. E. Davis, *Origins and Evolution*. Davis notes, however, a growing tendency toward "municipalization" of CLTs, in which local governments have become involved in CLTs or even initiated them.

31. J. E. Davis, *Origins and Evolution*; Daniel, *Dispossession*; Gilbert, Sharp, and Felin, "Loss and Persistence."

32. J. E. Davis, *Origins and Evolution*; Gray, "Community Land Trusts"; Penniman, *Farming While Black*, 13.

33. High land prices and insecure tenure are among the greatest obstacles faced by community gardens, so CLT efforts to create affordable and sustainable land access are a welcome development for the urban-agriculture movement. In a 2012 survey by the National Community Land Trust Network, thirty-seven CLTs from all over the US reported encouraging some kind of rural or urban agricultural activities (Rosenberg and Yuen, "Beyond Housing").

34. Tharpar, "Future of Farmland (Part 2)"; South of the Sound Community Farm Land Trust, "Community Farm Land Trust"; Sustainable Iowa Land Trust, "SILT."

35. Duke and Lynch, "Gauging Support"; Merenlender et al., "Land Trusts and Conservation Easements"; Wittman, Dennis, and Pritchard, "Beyond the Market?" For a comprehensive consideration of alternative approaches to landownership see Geisler and Daneker, *Property and Values*.

36. LaVecchia, "These Neighbors Got Together"; Margolin, "This Group Wants to Help Lower Rents."

37. Tharpar, "Future of Farmland (Part 2)"; Black Land Matters Real Estate Investment Cooperative, "Members Overview."

38. LaVecchia, "These Neighbors Got Together."

39. Block, *Postindustrial Possibilities*.

40. Polanyi, *Great Transformation*.

41. See Martin and Clapp, "Finance for Agriculture."

42. See, for example, Fairbairn, "Indirect Dispossession."

43. GRAIN, "Land Ceilings."

44. George, *Progress and Poverty*, bk. 8, chap. 2, para. 12, original emphasis.

45. *Economist*, "Why Land Value Taxes Are So Popular."

46. Popper, "Restyled as Real Estate Trusts."

47. Reydon et al., "Preço Elevado e o ITR."

48. J. E. Davis, *Origins and Evolution*.

49. Austenå, "Agrarian Land Law in Norway"; Kajii, "Development of Structural Policy"; National Agricultural Law Center, "Corporate Farming Laws"; Schutz, "Corporate-Farming Measures"; Welsh, Carpentier, and Hubbell, "On the Effectiveness." There is reason to believe that such policies are actually effective: Desmarais et al. ("Investor Ownership") found that investor ownership of farmland in Saskatchewan Province, Canada, increased sixteen-fold in the twelve years after a law restricting ownership to province residents was lifted in 2002.

50. Comissão Pastoral da Terra, "Campanha Nacional pelo Limite da Propriedade da Terra"; interview with campaign organizer. The limit proposed by the campaign was 35 MEI, which varies between 175 and 1,750 ha in different parts of the country.

# Bibliography

Abolafia, Mitchel. *Making Markets: Opportunism and Restraint on Wall Street.* Cambridge, MA: Harvard University Press, 1996.

Abraf. "Relatório de impactos negativos do parecer AGU/CGU LA 1/2008 ao setor de florestas plantadas e apresentação de sugestões a serem atendidas visando a superação dos impasses surgidos." Brasília: Associação Braileira de Produtores de Florestas Plantadas, 2010.

AcreTrader. "How it Works." https://www.acretrader.com/resources/how-it-works.

Adecoagro. "Institutional Presentation." 2018. https://ir.adecoagro.com/uploads/AGRO-IR-presentation.pdf.

———. "Organizational Structure." 2019. https://ir.adecoagro.com/index.php?s=organizational-structure.

Agrifirma Brazil. "$82m Investment by Private Equity Firm to Accelerate Farmland Transformation in Brazil." Press release, September 5, 2011. https://landmatrix.org/media/uploads/press_release_05_sept_2011.pdf.

Agrolink. "Discreta, Tiba Agro investe pesado na aquisição de terras." March 15, 2010. https://www.agrolink.com.br/noticias/discreta--tiba-agro-investe-pesado-na-aquisicao-de-terras_106924.html.

Allan, J. A. "Virtual Water: A Strategic Resource, Global Solutions to Regional Deficits." *Groundwater* 36, no. 4 (1998): 545–46.

Alston, Lee, Gary Libecap, and Bernardo Mueller. *Titles, Conflict, and Land Use: The Development of Property Rights and Land Reform on the Brazilian Amazon Frontier.* Ann Arbor: University of Michigan Press, 1999.

Alvim, Augusto Mussi. "Investimentos estrangeiros diretos e suas relações com os processos, causas e efeitos da concentração e estrangeirização das terras no Brasil." Brasília: Núcleo de Estudos Agrários e Desenvolvimento Rural (NEAD), 2009.

Amann, Edmund, and Werner Baer. "Neoliberalism and Its Consequences in Brazil." *Journal of Latin American Studies* 34, no. 4 (2002): 945–59.

Anderson, Benjamin. *Social Value: A Study in Economic Theory, Critical and Constructive.* New York: Houghton Mifflin, 1911.

Anderson, Scott. "Real Estate Valuation Strategy for Farmland—Cap Rate." AgWeb, December 18, 2018. https://www.agweb.com/blog/cashcow-farmer/real-estate-valuation-strategy-for-farmland-cap-rate/.

Andrejevic, Mark. "Surveillance in the Digital Enclosure." *Communication Review* 10, no. 4 (2007): 295–317.

Ankersen, Thomas T., and Thomas Ruppert. "Tierra y Libertad: The Social Function Doctrine and Land Reform in Latin America." *Tulane Environmental Law Journal* 19 (2006): 69.Arendell, Terry. "Reflections on the Researcher-Researched Relationship: A Woman Interviewing Men." *Qualitative Sociology* 20, no. 3 (1997): 341–68.

Arrighi, Giovanni. *The Long Twentieth Century: Money, Power, and Origins of Our Times.* New York: Verso Books, 1994.

Austenå, Torgeir. "Agrarian Land Law in Norway." In *Agrarian Land Law in the Western World,* edited by Margaret Rosso Grossman and Wim Brussaard, 134–51. Wallingford, UK: C.A.B. International, 1992.

Austin, J. L. *How to Do Things with Words*. 2nd ed. Cambridge, MA: Harvard University Press, 1975.

Avis, Ed. "Private Equity Money and Grocery Retailers: What Does It Mean?" *Retail Leader*, January 1, 2015. https://retailleader.com/private-equity-money-and-grocery-retailers-what-does-it-mean.

Babb, Sarah. "A Sociologist among Economists: Some Thoughts on Methods, Positionality, and Subjectivity." In *Emergent Methods in Social Research*, edited by Sharlene Hesse-Biber and Patricia Leavy, 43–53. Thousand Oaks, CA: Sage, 2006.

Bair, Jennifer. "Global Capitalism and Commodity Chains: Looking Back, Going Forward." *Competition and Change* 9, no. 2 (2005): 153–80.

Baka, Jennifer. "The Political Construction of Wasteland: Governmentality, Land Acquisition and Social Inequality in South India." *Development and Change* 44, no. 2 (2013): 409–28.

Bakker, Karen. "Katrina: The Public Transcript of 'Disaster.'" *Environment and Planning D* 23 (2005): 795–802.

Barkema, Alan D. "Farmland Values: The Rise, the Fall, the Future." *Economic Review* 72, no. 4 (April 1987): 19–35.

Barnett, Barry J. "The US Farm Financial Crisis of the 1980s." *Agricultural History* 74, no. 2 (2000): 366–80.

Beck, Ulrich. *Risk Society: Towards a New Modernity*. London: Sage, 1992.

Beckert, Jens. *Imagined Futures: Fictional Expectations and Capitalist Dynamics*. Cambridge, MA: Harvard University Press, 2016.

Berger, Sebastien. "Madagascar's New Leader Cancels Korean Land Deal." *Telegraph*, March 18, 2009. http://www.telegraph.co.uk/news/worldnews/africaandindianocean/madagascar/5012961/Madagascars-new-leader-cancels-Korean-land-deal.html.

Binkley, Clark S., Charles F. Raper, and Cortland L. Washburn. "Institutional Ownership of US Timberland." *Journal of Forestry* 94, no. 9 (1996): 21–28.

Bjorhus, Jennifer. "Hot Money Turns from Stocks to Farmland." *Star Tribune* (Minneapolis), July 5, 2012. http://www.startribune.com/part-2-hot-investments-turn-from-the-stock-market-to-farmland/160140665/.

Bjørkhaug, Hilde, André Magnan, and Geoffrey Lawrence, eds. *The Financialization of Agri-food Systems*. London: Earthscan, 2018.

Black Land Matters Real Estate Investment Cooperative (BLM-REIC). "Members Overview." *Black Land Matters*. http://www.duendenatural.com/blm/assets/black-land-matters-real-estate-investment-cooperative-members-overview.pdf.

Bleiberg, Robert. "Country Slickers: The Hue and Cry over Ag-Land Is Groundless." *Barron's National Business and Financial Weekly*, February 28, 1977.

Block, Fred. *Postindustrial Possibilities: A Critique of Economic Discourse*. Berkeley: University of California Press, 1990.

Blomley, Nicholas. "Mud for the Land." *Public Culture* 14, no. 3 (2002): 557–82.

Bobenrieth, Eugenio, and Brian Wright. "The Food Price Crisis of 2007/2008: Evidence and Implications." Food and Agriculture Organization of the United Nations (FAO), 2009. http://www.fao.org/fileadmin/templates/est/meetings/joint_igg_grains/Panel_Discussion_paper_2_English_only.pdf.

Bonnefield. "Four Sons and a Farm." June 14, 2017. https://bonnefield.com/pf/four-sons-and-a-farm/.

Borras, Saturnino, Jr., and Jennifer Franco. "From Threat to Opportunity? Problems with the Idea of a Code of Conduct for Land-Grabbing." *Yale Human Rights and Development Law Journal* 13 (2010): 507–23.

Borras, Saturnino, Jr., Jennifer Franco, Sergio Gómez, Cristóbal Kay, and Max Spoor. "Land Grabbing in Latin America and the Caribbean." *Journal of Peasant Studies* 39, no. 3–4 (2012): 845–72.

Boyd, William, Scott Prudham, and Rachel Schurman. "Industrial Dynamics and the Problem of Nature." *Society and Natural Resources* 14 (2001): 555–70.

Brandão, Paulo. "A formação territorial do Oeste Baiano: A constituição do 'Além São Francisco' (1827–1985)." *GeoTextos* 6, no. 1 (2010): 35–50.

Brannstrom, Christian. "Producing Possession: Labour, Law and Land on a Brazilian Agricultural Frontier, 1920–1945." *Political Geography* 20, no. 7 (2001): 859–83.

BrasilAgro. "Institutional Presentation." 2017. http://www.brasil-agro.com/brasilagro2011/web/default_en.asp?idioma=1&conta=44.

*Brasil de Fato.* "Assentamentos: Temer leva um ano para fazer o que Lula e Dilma fizeram em 10 dias." August 19, 2018. https://www.brasildefato.com.br/2018/08/19/temer-faz-2-dos-assentamentos-que-fizeram-lula-e-dilma/.

Braun, Bruce, and James McCarthy. "Hurricane Katrina and Abandoned Being." *Environment and Planning D* 23 (2005): 802–9.

Brautigan, Deborah. *Will Africa Feed China?* Oxford: Oxford University Press, 2015.

Brazilian Senate. "Ata da 4a reunião (conjunta) da Comissão de Agricultura e Reforma Agrária e 4a reunião da Comissão de Meio Ambiente, Defesa do Consumidor e Fiscalização e Controle." Diário do Senado Federal-Suplemento, May 3, 2008.

Brazil Travel. Map of Biomes of Brazil. Wikimedia Commons, January 1, 2015. https://commons.wikimedia.org/wiki/File:Biomes_of_Brazil.png.

Breger Bush, Sasha. *Derivatives and Development: A Political Economy of Global Finance, Farming, and Poverty.* New York: Palgrave Macmillan, 2012.

Brenner, Neil, and Nik Theodore. "Cities and the Geographies of 'Actually Existing Neoliberalism.'" *Antipode* 34, no. 3 (2002): 349–79.

Bronson, Kelly, and Irena Knezevic. "Big Data in Food and Agriculture." *Big Data and Society*, January–June 2016, 1–5.

Brown, J. Christopher, and Mark Purcell. "There's Nothing Inherent about Scale: Political Ecology, the Local Trap, and the Politics of Development in the Brazilian Amazon." *Geoforum* 36, no. 5 (2005): 607–24. doi:10.1016/j.geoforum.2004.09.001.

Burch, David, and Geoffrey Lawrence. "Financialization in Agri-food Supply Chains: Private Equity and the Transformation of the Retail Sector." *Agriculture and Human Values* 30, no. 2 (2013): 247–58.

——. "Towards a Third Food Regime: Behind the Transformation." *Agriculture and Human Values* 26, no. 4 (2009): 267–79.

Burgis, Tom. "The Great Land Rush: The Billionaire's Farm in Ethiopia." *Financial Times*, March 1, 2016. https://ig.ft.com/sites/land-rush-investment/ethiopia/?mhq5j=e7.

——. "Madagascar Scraps Daewoo Farm Deal." *Financial Times*, March 18, 2009. https://www.ft.com/content/7e133310-13ba-11de-9e32-0000779fd2ac.

Butler, Judith. "Performative Agency." *Journal of Cultural Economy* 3, no. 2 (2010): 147–61.

Caetano, Marcela. "Porteira aberta para os estrangeiros." *Dinheiro Rural*, May 30, 2017. https://www.dinheirorural.com.br/porteira-aberta-para-os-estrangeiros/.

Callon, Michel. "The Embeddedness of Economic Markets in Economics." In *The Laws of the Markets*, edited by Michel Callon, 1–57. Oxford: Oxford University Press, 2007.

——. "Performativity, Misfires and Politics." *Journal of Cultural Economy* 3, no. 2 (2010): 163–69.

——. "What Does It Mean to Say That Economics Is Performative?" In *Do Economists Make Markets? On the Performativity of Economics*, edited by Donald MacKenzie, Fabian Muniesa, and Lucia Siu, 311–57. Princeton, NJ: Princeton University Press, 1998.

Câmara dos Deputados. "Relatório da subcomissão destinada a, no prazo de 180 dias, analisar e propor medidas sobre o processo de aquisição de áreas rurais e suas utilizações, no Brasil, por pessoas físicas e jurídicas estrangeiras." Brasília: Subestra, 2011.

Cardoso, Fernando Henrique, and Enzo Faletto. *Dependency and Development in Latin America.* Berkeley: University of California Press, 1979.

Carolan, Michael. "Barriers to the Adoption of Sustainable Agriculture on Rented Land: An Examination of Contesting Social Fields." *Rural Sociology* 70, no. 3 (2005): 387–413.

——. "Publicising Food: Big Data, Precision Agriculture, and Co-experimental Techniques of Addition." *Sociologia Ruralis* 57, no. 2 (2017): 135–54.

Carter, Miguel. "Broken Promise: The Land Reform Debacle under the PT Governments." In *Challenging Social Inequality: The Landless Rural Workers Movement and Agrarian Reform in Brazil*, edited by Miguel Carter, 413–28. Durham, NC: Duke University Press, 2015.

Cass, David, and Karl Shell. "Do Sunspots Matter?" *Journal of Political Economy* 91, no. 2 (1983): 193–227.

Chong, Kimberly, and David Tuckett. "Constructing Conviction through Action and Narrative: How Money Managers Manage Uncertainty and the Consequence for Financial Market Functioning." *Socio-Economic Review* 13, no. 2 (2005): 309–30.

Christophers, Brett. "For Real: Land as Capital and Commodity." *Transactions of the Institute of British Geographers* 41, no. 2 (2016): 134–48.

——. "The Limits to Financialization." *Dialogues in Human Geography* 5, no. 2 (2015): 183–200.

——. "On Voodoo Economics: Theorising Relations of Property, Value and Contemporary Capitalism." *Transactions of the Institute of British Geographers* 35, no. 1 (2010): 94–108.

Chryst, Walter. "Land Values and Agricultural Income: A Paradox?" *Journal of Farm Economics* 47, no. 5 (1965): 1265–73.

CiBO Technologies. "About." https://www.cibotechnologies.com/about/.

Clapp, Jennifer. "Financialization, Distance and Global Food Politics." *Journal of Peasant Studies* 41, no. 5 (2014): 797–814.

——. *Food.* Cambridge: Polity, 2012.

——. "Responsibility to the Rescue? Governing Private Financial Investment in Global Agriculture." *Agriculture and Human Values* 34 (2017): 223–35.

Clapp, Jennifer, and Eric Helleiner. "Troubled Futures? The Global Food Crisis and the Politics of Agricultural Derivatives Regulation." *Review of International Political Economy* 19, no. 2 (2012): 181–207.

Clapp, Jennifer, and Ryan Isakson. *Speculative Harvests: Financialization, Food, and Agriculture.* Halifax, NS: Fernwood, 2018.

Clark, Gordon. "Pension Fund Capitalism: A Causal Analysis." *Geografiska Annaler* 80, no. 3 (1998): 139–57.

Cole, Robert. "The New Black Gold: U.S. Farmland." *Globe and Mail*, March 22, 2012, B11.

Colman, Greg, and Jeff Daley. "Global Farmland Primer: Investment Opportunities in Farmland." Wellington West Capital Markets, August 21, 2008. http://www.farmlandinvestmentpartnership.com/wp-content/uploads/2014/08/128_GlobalFarmlandPrimer082108.pdf (site discontinued).

Comissão Pastoral da Terra (CPT). "Campanha Nacional pelo Limite da Propriedade da Terra." May 13, 2010. https://www.cptnacional.org.br/acoes/campanhas/campanha-pelo-limite-da-propriedade-da-terra/210-campanha-nacional-pelo-limite-da-propriedade-da-terra.

Committee on Food Security. *Voluntary Guidelines on the Responsible Governance of Tenure of Land, Fisheries and Forests in the Context of National Food Security.* Rome, 2012. http://www.fao.org/docrep/016/i2801e/i2801e.pdf.

Conrad, Jeff. "Expert Commentary: A Decade in the Farmland Asset Class Evolution." Global AgInvesting. http://www.globalaginvesting.com/decade-farmland-asset-class-evolution/.

Cortese, Amy. "The Crowdfunding Crowd Is Anxious." *New York Times*, January 5, 2013. http://www.nytimes.com/2013/01/06/business/crowdfunding-for-small-business-is-still-an-unclear-path.html.

Crippen, Alex. "CNBC Buffett Transcript Part 2: The 'Zebra' That Got Away." CNBC.com, March 2, 2011. https://www.cnbc.com/id/41867379.

Crittenden, Ann. "Farmland Lures Institutions." *New York Times*, November 24, 1980.

Cronon, William. *Nature's Metropolis: Chicago and the Great West.* New York: W. W. Norton, 1992.

Cunha, Alexandre dos Santos. "The Social Function of Property in Brazilian Law." *Fordham Law Review* 80, no. 3 (2011): 1171–81.

Daniel, Pete. *Dispossession: Discrimination against African American Farmers in the Age of Civil Rights.* Chapel Hill: University of North Carolina Press, 2013.

Davis, Gerald. *Managed by Markets: How Finance Re-shaped America.* Oxford: Oxford University Press, 2009.

Davis, John Emmeus. *Origins and Evolution of the Community Land Trust in the United States.* Lincoln Institute of Land Policy, 2014. https://community-wealth.org/sites/clone.community-wealth.org/files/downloads/report-davis14.pdf.

DeBraal, J. Peter, and Kenneth Krause. "Corporate, Foreign, and Financial Investors in U.S. Agriculture." *South Dakota Law Review* 29 (1983): 378–427.

De Goede, Marieke. "Discourses of Scientific Finance and the Failure of Long-Term Capital Management." *New Political Economy* 6, no. 2 (2001): 149–70.

——. *Virtue, Fortune, and Faith: A Genealogy of Finance.* Minneapolis: University of Minnesota Press, 2005.

Deininger, Klaus, and Derek Byerlee. *Rising Global Interest in Farmland: Can It Yield Sustainable and Equitable Benefits?* Washington, DC: World Bank, 2011.

Delgado, Guilherme. "A questão agrária no Brasil, 1950–2003." In *Questão social e políticas sociais no Brasil contemporâneo*, compiled by Luciana Jaccoud, 51–90. Brasília: Instituto de Pesquisa Econômica Aplicada (IPEA), 2005.

——. *Capital financeiro e agricultura no Brasil: 1965–1985.* São Paulo: Ícone Editora Ltda, 1985.

De Sartre, Xavier, and Romain Taravella. "National Sovereignty vs. Sustainable Development: Lessons from the Narrative on the Internationalization of the Brazilian Amazon." *Political Geography* 28 (2009): 406–15.

De Schutter, Olivier. "How Not to Think about Land-Grabbing: Three Critiques of Large-Scale Investments in Farmland." *Journal of Peasant Studies* 38, no. 2 (2011): 249–79.

Desmarais, Annette, Darrin Qualman, André Magnan, and Nettie Wiebe. "Investor Ownership or Social Investment? Changing Farmland Ownership in Saskatchewan, Canada." *Agriculture and Human Values* 34, no. 1 (2017): 149–66.

De Sousa, Ivan, and Lawrence Busch. "Networks and Agricultural Development: The Case of Soybean Production and Consumption in Brazil." *Rural Sociology* 63, no. 3 (1998): 349–71.

Dias, Guilherme Leite da Silva, Claudio Afonso Vieira, and Cicely Moitinho Amaral. *Comportamento do mercado de terras no Brasil*. Santiago, Chile: CEPAL, 2001.

Dodson, Don. "Ag Real Estate Firm Cultivating Opportunities." *News-Gazette* (Champaign, IL), July 1, 2007. http://www.news-gazette.com/news/local/2007-07-01/ag-real-estate-firm-cultivating-opportunities.html.

Dotzour, Mark. "Peoples Company Land Expo Keynote Speaker Dr. Mark Dotzour." January 25, 2012. https://youtu.be/41lzJqt5xiA.

Ducastel, Antoine, and Ward Anseeuw. "Agriculture as an Asset Class: Reshaping the South African Farming Sector." *Agriculture and Human Values* 34 (2017): 199–209.

Dudley, Kathryn Marie. *Debt and Dispossession: Farm Loss in America's Heartland*. Chicago: University of Chicago Press, 2000.

Duke, Joshua, and Lori Lynch. "Gauging Support for Innovative Farmland Preservation Techniques." *Policy Sciences* 40, no. 2 (2007): 123–55.

Duménil, Gérard, and Dominique Lévy. "Costs and Benefits of Neoliberalism: A Class Analysis." In *Financialization and the World Economy*, edited by Gerald Epstein, 17–45. Cheltenham, UK: Edward Elgar, 2005.

Dunkley, Dan. "Investors Are Going Nuts for Nuts." *Wall Street Journal*, February 11, 2014. https://blogs.wsj.com/privateequity/2014/02/11/investors-are-going-nuts-for-nuts/.

*Economist*. "Bonds That Rock and Roll." May 7, 1998. http://www.economist.com/node/127644.

——. "When Others Are Grabbing Their Land." May 5, 2011. https://www.economist.com/international/2011/05/05/when-others-are-grabbing-their-land.

——. "Why Land Value Taxes Are So Popular, Yet So Rare." November 10, 2014. https://www.economist.com/blogs/economist-explains/2014/11/economist-explains-0.

Edelman, Marc. "Messy Hectares: Questions about the Epistemology of Land Grabbing Data." *Journal of Peasant Studies* 40, no. 3 (2013): 485–501.

Ekers, Michael. "Financiers in the Forests on Vancouver Island, British Columbia: On Fixes and Colonial Enclosures." *Journal of Agrarian Change* 19, no. 2 (April 2019): 270–94.

Epstein, Gerald A. "Introduction: Financialization and the World Economy." In *Financialization and the World Economy*, edited by Gerald Epstein, 3–16. Cheltenham, UK: Edward Elgar, 2005.

Espeland, Wendy. "Value-Matters." *Economic and Political Weekly* 36, no. 21 (2001): 1839–45.

Espeland, Wendy, and Paul Hirsch. "Ownership Changes, Accounting Practice and the Redefinition of the Corporation." *Accounting, Organizations and Society* 15, no. 1/2 (1990): 77–96.

Espeland, Wendy, and Mitchell Stevens. "Commensuration as a Social Process." *Annual Review of Sociology* 24 (1998): 313–43.

Esposito, Elena. *The Future of Futures: The Time of Money in Financing and Society*. Translated by Elena Esposito and Andrew Whitehead. Cheltenham, UK: Edward Elgar, 2011.

*Estado de São Paulo*. "Mataram Selig, o que vendia nossas terras." March 12, 1970, 7.

Evans, Peter. *Dependent Development: The Alliance of Multinational, State, and Local Capital in Brazil*. Princeton, NJ: Princeton University Press, 1979.

Fairbairn, Madeleine. "Foreignization, Financialization, and Land Grab Regulation." *Journal of Agrarian Change* 15, no. 4 (2015): 581–91.

——. "Indirect Dispossession: Domestic Power Imbalances and Foreign Access to Land in Mozambique." *Development and Change* 44, no. 2 (2013): 335–56.

——. "'Like Gold with Yield': Evolving Intersections between Farmland and Finance." *Journal of Peasant Studies* 41, no. 5 (2014): 777–95.

——. "Reinventing the Wheel? Or Adding New Air to Old Tires?" *Dialogues in Human Geography* 5, no. 2 (2015): 183–200.

Fairhead, James, Melissa Leach, and Ian Scoones. "Green Grabbing: A New Appropriation of Nature?" *Journal of Peasant Studies* 39, no. 2 (2012): 237–61.

Falk, Barry, and Bong-Soo Lee. "Fads versus Fundamentals in Farmland Prices." *American Journal of Agricultural Economics* 80, no. 4 (1998): 696–707.

FAO, IFAD, UNCTAD, and World Bank Group. "Principles for Responsible Agricultural Investment That Respects Rights, Livelihoods and Resources." Washington, DC: World Bank, January 25, 2010. http://siteresources.worldbank.org/INT ARD/214574-1111138388661/22453321/Principles_Extended.pdf.

Farmland LP. Home page. http://www.farmlandlp.com/.

Farmland Partners. "Farmland Partners Files Lawsuit against Rota Fortunae and Co-conspirators for 'Short and Distort' Attack." Press release, July 24, 2018. https://www.prnewswire.com/news-releases/farmland-partners-files-lawsuit-against-rota-fortunae-and-co-conspirators-for-short-and-distort-attack-300685269.html.

——. *2016 Annual Report to Stockholders*. 2017. http://ir.farmlandpartners. com/Cache/1001221793.PDF?Y=&O=PDF&D=&fid=1001221793&T=& iid=4426904.

Featherstone, Allen, and Charles Moss. "Capital Markets, Land Values, and Boom-Bust Cycles." In *Government Policy and Farmland Markets: The Maintenance of Farmer Wealth*, edited by Charles Moss and Andrew Schmitz, 159–78. Ames: Iowa State Press, 2003.

Fikade, Birhanu. "Karuturi to Start Afresh in Ethiopia." *Reporter* (Addis Ababa), April 21, 2018. https://www.thereporterethiopia.com/article/karuturi-start-afresh-ethiopia.

Fischer, Gunther, and Mahendra Shah. "Farmland Investments and Food Security: Statistical Annex." Washington, DC: World Bank, 2010. http://documents. worldbank.org/curated/en/884731468221080363/Farmland-investments-and-food-security.

Fisher, Monica, and Vongai Kandiwa. "Can Agricultural Input Subsidies Reduce the Gender Gap in Modern Maize Adoption? Evidence from Malawi." *Food Policy* 45 (2014): 101–11.

Fitzgerald, Deborah. *Every Farm a Factory: The Industrial Ideal in American Agriculture*. New Haven, CT: Yale University Press, 2003.

FitzSimmons, Margaret. "The New Industrial Agriculture." *Economic Geography* 62, no. 4 (1986): 334–53.

Fligstein, Neil. *The Architecture of Markets: An Economic Sociology of Twenty-First-Century Capitalist Societies*. Princeton, NJ: Princeton University Press, 2001.

——. "The End of (Shareholder Value) Ideology?" *Political Power and Social Theory* 17 (2005): 223–28.

Flint, Jerry. "Solid Gold Down on the Farm." *New York Times*, April 24, 1977.

Foley, Duncan. *Adam's Fallacy: A Guide to Economic Theology*. Cambridge, MA: Belknap Press of Harvard University Press, 2006.

Food and Agriculture Organization of the United Nations. "World Food Situation." http://www.fao.org/worldfoodsituation/foodpricesindex/en/.

*Forbes.* "Land Anyone?" February 1, 1977.

Fourcade, Marion. "Cents and Sensibility: Economic Valuation and the Nature of 'Nature.'" *American Journal of Sociology* 116, no. 6 (2011): 1721–77.

Frank, Andre Gunder. *Capitalism and Underdevelopment in Latin America: Historical Studies of Chile and Brazil.* New York: Monthly Review, 1969.

Friedland, William, Amy Barton, and Robert Thomas. *Manufacturing Green Gold: Capital, Labor and Technology in the Lettuce Industry.* Cambridge: Cambridge University Press, 1981.

Friedman, Milton. "A Friedman Doctrine: The Social Responsibility of Business Is to Increase Its Profits." *New York Times Magazine,* September 13, 1970.

Friedmann, Harriet. "The Political Economy of Food: A Global Crisis." *New Left Review* 197 (1993): 29–57.

Fritz, Michael. "As Cash Backlog Builds, Farmland Fund Paying $108 Million for 29,000 Acres in Texas and Wisconsin." Farmland Investor Center, December 3, 2012. http://www.farmlandinvestorcenter.com/?p=302&option=com_wordpress&Itemid.

——. "Illinois Pension Board, an Early Adopter in Farmland Investing, to Exit the Sector." Farmland Investor Center, December 19, 2014. http://www.farmlandinvestorcenter.com/?p=1198&option=com_wordpress&Itemid=171.

——. "Institutional Investment in US Farmland." Speech delivered at 2012 Ag Summit, Chicago, December 11, 2012. http://www.dtnprogressivefarmer.com/go/agsummit/agsummit2012files/presentations/Working%20with%20Farm%20Investors.Fritz.pdf.

Froud, Julie, Colin Haslam, Sukhdev Johal, and Karel Williams. "Shareholder Value and Financialization: Consultancy Promises, Management Moves." *Economy and Society* 29, no. 1 (2000): 80–110.

Froud, Julie, and Karel Williams. "Private Equity and the Culture of Value Extraction." *New Political Economy* 12, no. 3 (2007): 405–20.

Fundação Getulio Vargas Instituto Brasileiro de Economia (FGV IBRE). "FGV Dados." http://portalibre.fgv.br/main.jsp?lumChannelId=402880811D8E2C4C011D8E33F5700158#.

Gage, G. "Farmland Partners Helps Stabilize Agriculture for Farmers, Investors." *REIT Magazine,* July/August 2018. https://www.reit.com/news/reit-magazine/july-august-2018.

Garrido Filha, Irene. *O Projeto Jari e os capitais estrangeiros na Amazônia.* Petrópolis, Brazil: Vozes, 1980.

Garud, Raghu. "Conferences as Venues for the Configuration of Emerging Organizational Fields: The Case of Cochlear Implants." *Journal of Management Studies* 45, no. 6 (2008): 1061–88.

Geisler, Charles. "New Terra Nullius Narratives and the Gentrification of Africa's 'Empty Lands.'" *Journal of World-Systems Research* 18, no. 1 (2012): 15–29.

Geisler, Charles, and Gail Daneker, eds. *Property and Values: Alternatives to Public and Private Ownership.* Washington, DC: Island Press, 2000.

George, Henry. *Progress and Poverty: An Inquiry into the Cause of Industrial Depressions and of Increase of Want with Increase of Wealth: The Remedy.* Garden City, NY: Doubleday, Page & Co., 1912. First published in 1879. http://www.econlib.org/library/YPDBooks/George/grgPPCover.html.

Gereffi, Gary. "The Organization of Buyer-Driven Global Commodity Chains: How U.S. Retailers Shape Overseas Production Networks." In *Commodity Chains and Global Capitalism,* edited by Gary Gereffi and Miguel Korzeniewicz, 95–122. Westport, CT: Praeger, 1994.

Gieryn, Thomas F. "Boundary-Work and the Demarcation of Science from Non-science: Strains and Interests in Professional Ideologies of Scientists." *American Sociological Review* 48, no. 6 (1983): 781–95.

Gilbert, Jess, Gwen Sharp, and M. Sindy Felin. "The Loss and Persistence of Black-Owned Farms and Farmland: A Review of the Research Literature and Its Implications." *Southern Rural Sociology* 18, no. 2 (2002): 1–30.

Gilbert, Katie. "Dirt Rich?" *Institutional Investor*, April 2010.

Gladstone Land. *2016 Annual Report*. 2017. https://ir.gladstoneland.com/static-files/829daf84-7012-4f27-8d49-695cd3aec677.

Gliessman, Stephen, and Eric Engles. *Agroecology: The Ecology of Sustainable Food Systems*. 3rd ed. Boca Raton, FL: CRC, 2015.

Gloy, Brent, Chris Hurt, Michael Boehlje, and Craig Dobbins. "Farmland Values: Current and Future Prospects." Department of Agricultural Economics, Purdue University, 2011.

Goldman, Michael. *Imperial Nature: The World Bank and Struggles for Social Justice in the Age of Globalization*. New Haven, CT: Yale University Press, 2005.

Gómez-Baggethun, Erik, Rudolf de Groot, Pedro L. Lomas, and Carlos Montes. "The History of Ecosystem Services in Economic Theory and Practice: From Early Notions to Markets and Payment Schemes." *Ecological Economics* 69, no. 6 (2010): 1209–18.

Goodman, David, Bernardo Sorj, and John Wilkinson. *From Farming to Biotechnology: A Theory of Agro-Industrial Development*. Oxford: Basil Blackwell, 1987.

Graeber, David. "Value: Anthropological Theories of Value." In *A Handbook of Economic Anthropology*, edited by James Carrier, 439–54. Cheltenham, UK: Edward Elgar, 2005.

GRAIN. "Land Ceilings: Reining in Land Grabbers or Dumbing Down the Debate?" *Against the Grain*, February 2013. http://www.grain.org/article/entries/4655-land-ceilings-reining-in-land-grabbers-or-dumbing-down-the-debate.

——. "Responsible Farmland Investing? Current Efforts to Regulate Land Grabs Will Make Things Worse." *Against the Grain*, August 2012. https://www.grain.org/article/entries/4564-responsible-farmland-investing-current-efforts-to-regulate-land-grabs-will-make-things-worse.

——. "Seized: The 2008 Land Grab for Food and Financial Security." Barcelona, October 24, 2008. http://www.grain.org/article/entries/93-seized-the-2008-landgrab-for-food-and-financial-security.

——. "Socially Responsible Farmland Investment: A Growing Trap." *Against the Grain*, October 2015. https://www.grain.org/article/entries/5294-socially-responsible-farmland-investment-a-growing-trap.

GRAIN and Rede Social de Justiça e Direitos Humanos. "Harvard's Billion-Dollar Farmland Fiasco." September 2018. https://www.grain.org/article/entries/6006-harvard-s-billion-dollar-farmland-fiasco.

Grantham, Jeremy. "Time to Wake Up: Days of Abundant Resources and Falling Prices Are Over Forever." GMO Quarterly Letter, April 2011. http://www.scribd.com/doc/54681895/Jeremy-Grantham-Investor-Letter-1Q-2011.

Grassroots International. "Tell TIAA to Go Deforestation and Land Grab-Free!" http://act.grassrootsonline.org/p/dia/action/public/?action_KEY=23765.

Gray, Karen. "Community Land Trusts in the United States." *Journal of Community Practice* 16, no. 1 (2008): 65–78.

Graziano da Silva, José. *A modernização dolorosa: Estrutura agrária, fronteira agrícola e trabalhadores rurais no Brasil*. Rio de Janeiro: Zahar Editores, 1981.

Greenberger, Michael. "Derivatives in the Crisis and Financial Reform." In *The Political Economy of Financial Crises*, edited by Martin Wolfson and Gerald Epstein, 467–90. Oxford: Oxford University Press, 2013.

Gunnoe, Andrew. "The Financialization of the US Forest Products Industry: Socioeconomic Relations, Shareholder Value, and the Restructuring of an Industry." *Social Forces* 94, no. 3 (2016): 1075–1101.

——. "The Political Economy of Institutional Landownership: Neorentier Society and the Financialization of Land." *Rural Sociology* 79, no. 4 (2014): 478–504.

Gunnoe, Andrew, and Paul K. Gellert. "Financialization, Shareholder Value, and the Transformation of Timberland Ownership in the US." *Critical Sociology* 37, no. 3 (2011): 265–84.

Gustke, Constance. "Digging into Farmland." CNBC, January 5, 2011. https://www.cnbc.com/id/40784051.

——. "Farm to Market: Taking Stock of the Agricultural Land Grab." CNBC, January 21, 2016.

Guthman, Julie. "Back to the Land: The Paradox of Organic Food Standards." *Environment and Planning A* 36 (2004): 511–28.

Hague, Fábio Santana, Marcus Peixoto, and José Vieira Filho. *Aquisição de terras por estrangeiros no Brasil: Uma avaliação jurídica e econômica.* Rio de Janeiro: Instituto de Pesquisa Econômica Aplicada (IPEA), 2012.

Haila, Anne. "The Theory of Land Rent at a Crossroads." *Environment and Planning D* 8, no. 3 (1990): 275–96.

Hall, Anthony. "Land Tenure and Land Reform in Brazil." In *Agrarian Reform and Grassroots Development: Ten Case Studies*, edited by Roy Prosterman, Mary Temple, and Timothy Hanstad, 205–32. London: Lynne Rienner, 1990.

Hall, Derek. *Land.* Cambridge: Polity, 2013.

Han, Jun, and Youguo Liang. "The Historical Performance of Real Estate Investment Trusts." *Journal of Real Estate Research* 10, no. 3 (1995): 235–62.

Hancock Agricultural Investment Group (HAIG). "About Us." http://hancockagriculture.com/about/.

——. "Farmland Management." http://hancockagriculture.com/farmland-management/.

——. "Hancock Agricultural Investment Group: A Company Overview." http://hancockagriculture.com/wp-content/uploads/sites/3/HAIG-Q318_final.pdf.

Harl, Neil. "Farm Corporations—Present and Proposed Restrictive Legislation." *Business Lawyer* 25 (1969): 1247–58.

Hartmann, Betsy. "The Ghosts of Malthus: Narratives and Mobilizations of Scarcity in the U.S. Political Context." In *The Limits to Scarcity: Contesting the Politics of Allocation*, edited by Lyla Mehta, 49–66. London: Earthscan, 2010.

Harvey, David. *The Enigma of Capital and the Crises of Capitalism.* Oxford: Oxford University Press, 2010.

——. "Globalization and the Spatial Fix." *Geographische Revue* 2 (2001): 23–30.

——. *The Limits to Capital.* London: Verso Books, 2006.

——. "The Urban Process under Capitalism: A Framework for Analysis." *International Journal of Urban and Regional Research* 2, no. 1–3 (1978): 101–31.

Headey, Derek, and Shenggen Fan. "Anatomy of a Crisis: The Causes and Consequences of Surging Food Prices." *Agricultural Economics* 39, no. s1 (2008): 375–91.

Hecht, Susanna. "Environment, Development and Politics: Capital Accumulation and the Livestock Sector in Eastern Amazonia." *World Development* 13, no. 6 (1985): 663–84.

Heilbroner, Robert. *The Worldly Philosophers: The Lives, Times, and Ideas of the Great Economic Thinkers.* 7th ed. New York: Simon & Schuster, 1999.

Hemphill, Peter. "Australian Farmland Index to Measure Corporate Farm Performance." *Weekly Times*, November 2, 2016. https://www.weeklytimesnow.com.au/agribusiness/australian-farmland-index-to-measure-corporate-farm-performance/news-story/281a6ef7185e679ff1352901751918cb.

Henderson, George. *California and the Fictions of Capital.* Philadelphia: Temple University Press, 1998.

Henderson, Jason. "Will Farmland Values Keep Booming?" *Economic Review* 93, no. 2 (2008): 81–104.

Henderson, Jason, Brent Gloy, and Michael Boehlje. "Agriculture's Boom-Bust Cycles: Is This Time Different?" *Economic Review* (Fourth Quarter 2011): 83–101.

Heppner, Kelvin. "Reasons to Sell Land and Lease It Back—a Toronto Investment Fund Manager's Pitch." *RealAgriculture*, November 25, 2014. https://www.realagriculture.com/2014/11/reasons-sell-land-lease-back-toronto-fund-managers-pitch/.

Hernandes, Tânia. "Estudo sobre processos, causas e efeitos da concentração e estrangeirização das terras no Brasil—estrutura de mercado." Brasília: Núcleo de Estudos Agrários e Desenvolvimento Rural (NEAD), 2009.

HighQuest Partners. "Private Financial Sector Investment in Farmland and Agricultural Infrastructure." Paris: Organisation for Economic Co-Operation and Development, 2010. http://dx.doi.org/10.1787/5km7nzpjlr8v-en.

HighQuest Partners and Julie Koeninger. "History of Institutional Farmland Investment." Danvers, MA: HighQuest Partners, 2017. http://www.globalaginvesting.com/wp-content/uploads/2017/04/Farmland_Investment_History_Koeninger_HQP.pdf.

Hildyard, Nicholas. "'Scarcity' as Political Strategy: Reflections on Three Hanging Children." In *The Limits to Scarcity: Contesting the Politics of Allocation*, edited by Lyla Mehta, 149–64. London: Earthscan, 2010.

Ho, Karen. *Liquidated: An Ethnography of Wall Street.* Durham, NC: Duke University Press, 2009.

Hodgson, Stephen, Cormac Cullinan, and Karen Campbell. *Land Ownership and Foreigners: A Comparative Analysis of Regulatory Approaches to the Acquisition and Use of Land by Foreigners.* Geneva: Food and Agriculture Organization of the UN, 1999.

Holston, James. "The Misrule of Law: Land and Usurpation in Brazil." *Comparative Studies in Society and History* 33, no. 4 (1991): 695–725.

Hoppe, Robert. *Structure and Finances of U.S. Farms: Family Farm Report, 2014 Edition.* EIB-132, US Department of Agriculture, Economic Research Service, December 2014.

Hubacek, Klaus, and Jeroen C. J. M. van den Bergh. "Changing Concepts of 'Land' in Economic Theory: From Single to Multi-disciplinary Approaches." *Ecological Economics* 56, no. 1 (2006): 5–27.

Hudson, Alan. "Beyond the Borders: Globalisation, Sovereignty and Extra-territoriality." *Geopolitics* 3, no. 1 (1998): 89–105.

Hull, John C. *Options, Futures, and Other Derivatives.* 8th ed. Saddle River, NJ: Prentice Hall, 2012.

Irwin, Scott H., and Dwight Sanders. "Index Funds, Financialization, and Commodity Futures Markets." *Applied Economic Perspectives and Policy* 33, no. 1 (2011): 1–31.

Isakson, S. Ryan. "Derivatives for Development? Small-Farmer Vulnerability and the Financialization of Climate Risk Management." *Journal of Agrarian Change* 15, no. 4 (2015): 569–80.

——. "Food and Finance: The Financial Transformation of Agro-food Supply Chains." *Journal of Peasant Studies* 41, no. 5 (2014): 749–75.

Jacobius, Arleen. "More Investors Turn to Farmland, Debt Strategies." *Pensions and Investments*, October 16, 2017. https://www.pionline.com/article/20171016/PRINT/171019921/more-investors-turn-to-farmland-debt-strategies.

Janiec, Chris. "Is Ag Suffering from Inflated Expectations?" Agri Investor, April 30, 2019. https://www.agriinvestor.com/inflated-expectations/.

——. "Peoples Company Launches Cover-Crop Initiative in Bid to Boost Land Value." Agri Investor, October 18, 2018. https://www.agriinvestor.com/peoples-company-launches-cover-crop-initiative-bid-boost-land-value/.

Jefferson, Thomas. "Chapter 15, Document 32: Thomas Jefferson to James Madison, 28 Oct. 1785." In *The Founders' Constitution*, edited by Philip Kurland and Ralph Lerner. Chicago: University of Chicago Press, 2000. http://press-pubs.uchicago.edu/founders/documents/v1ch15s32.html.

Jevons, William Stanley. "Commercial Crises and Sun-Spots." *Nature*, 1878.

Johnson, H. Thomas. "Postwar Optimism and the Rural Financial Crisis of the 1920s." *Explorations in Economic History* 11, no. 2 (1974): 173–92.

Johnson, Leigh. "Index Insurance and the Articulation of Risk-Bearing Subjects." *Environment and Planning A* 45 (2013): 2663–81.

Jordà, Òscar, Katharina Knoll, Dmitry Kuvshinov, Moritz Schularick, and Alan M. Taylor. "The Rate of Return on Everything, 1870–2015." Federal Reserve Bank of San Francisco Working Paper 2017–25, 2017. https://doi.org/10.24148/wp2017-25.

Jung-a, Song, Christian Oliver, and Tom Burgis. "Daewoo to Cultivate Madagascar Land for Free." *Financial Times*, November 19, 2008. https://www.ft.com/content/6e894c6a-b65c-11dd-89dd-0000779fd18c.

Kajii, Isoshi. "Development of Structural Policy: Centering around the Agricultural Land Legislation." In *Changes in Japan's Agrarian Structure*, edited by Isoshi Kajii, Shigeru Usami, and Sadako Nakayasu, 13–74. Tokyo: Food and Agriculture Policy Research Center, 1998.

Kaplan, Howard. "Farmland as a Portfolio Investment." *Journal of Portfolio Management* 11, no. 2 (Winter 1985): 73–78.

Kaufman, Frederick. *Bet the Farm: How Food Stopped Being Food*. Hoboken, NJ: John Wiley & Sons, 2012.

——. "The Food Bubble: How Wall Street Starved Millions and Got Away with It." *Harper's*, July 2010. http://frederickkaufman.typepad.com/files/the-food-bubble-pdf.pdf.

Kautsky, Karl. *The Agrarian Question*. Translated by Pete Burgess. London: Zwan, 1988.

Kay, Sylvia. "Political Brief on the Principles on Responsible Investment in Agriculture and Food Systems." Transnational Institute, March 2015. https://www.tni.org/en/briefing/political-brief-principles-responsible-investment-agriculture-and-foodsystems.

Kenney-Lazar, Miles. "Plantation Rubber, Land Grabbing and Social-Property Transformation in Southern Laos." *Journal of Peasant Studies* 39, no. 3–4 (2012): 1017–37.

Kindleberger, Charles, and Robert Aliber. *Manias, Panics, and Crashes: A History of Financial Crises*. Hoboken, NJ: John Wiley & Sons, 2005.

Kish, Zenia, and Madeleine Fairbairn. "Investing for Profit, Investing for Impact: Moral Performances in Agricultural Investment Projects." *Environment and Planning A* 50, no. 3 (2018): 569–88.

Klinenberg, Eric. "Denaturalizing Disaster: A Social Autopsy of the 1995 Chicago Heat Wave." *Theory and Society* 28, no. 2 (1999): 239–95.

Kloppenburg, Jack. *First the Seed: The Political Economy of Plant Biotechnology*. Madison: University of Wisconsin Press, 2004.

Knickmeyer, Ellen. "In Egypt, Upper Crust Gets the Bread." *Washington Post*, April 5, 2008. http://www.washingtonpost.com/wp-dyn/content/article/2008/04/04/AR2008040403937.html.

Knight, Frank. *Risk, Uncertainty and Profit*. Chicago: Hart, Schaffner and Marx, 1921.

Knuth, Sarah. "Global Finance and the Land Grab: Mapping Twenty-First Century Strategies." *Canadian Journal of Development Studies* 36, no. 2 (2015): 163–78.

Koeninger, Julie. "A Farmland Investment Primer." GMO White Paper, July 2014. https://www.valuewalk.com/wp-content/uploads/2014/07/JK_IntroTo Farmland_714-1.pdf.

Kolesnikova, Maria. "Grantham Says Farmland Will Outperform All Global Assets." *Bloomberg*, October 8, 2011. http://www.bloomberg.com/news/2011-08-10/grantham-says-farmland-will-outperform-all-global-assets-1-.html.

Kotz, David. "Financialization and Neoliberalism." In *Relations of Global Power: Neoliberal Order and Disorder*, edited by Gary Teeple and Stephen McBride, 1–18. Toronto: University of Toronto Press, 2011.

Koven, Peter. "ETF May Stand for Exchange-Traded Farmland." *Financial Post*, January 19, 2012. http://business.financialpost.com/2012/01/19/etf-may-stand-for-exchange-traded-farmland/.

Krippner, Greta. *Capitalizing on Crisis: The Political Origins of the Rise of Finance*. Cambridge, MA: Harvard University Press, 2011.

——. "The Financialization of the American Economy." *Socio-Economic Review* 3, no. 2 (2005): 173–208.

Kuns, Brian, Oane Visser, and Anders Wastfelt. "The Stock Market and the Steppe: The Challenges Faced by Stock-Market Financed, Nordic Farming Ventures in Russia and Ukraine." *Journal of Rural Studies* 45 (2016): 199–217.

Lacey, Marc. "Across Globe, Empty Bellies Bring Rising Anger." *New York Times*, April 18, 2008. http://www.nytimes.com/2008/04/18/world/americas/18food.html.

Land Commodities. *The Land Commodities Global Agriculture and Farmland Investment Report 2009*. Baar, Switzerland, 2009. http://www.farmlandinvestmentreport.com/Farmland_Investment_Report.pdf.

Langley, Paul. *The Everyday Life of Global Finance*. Oxford: Oxford University Press, 2008.

Lappé, Frances Moore. "Beyond the Scarcity Scare: Reframing the Discourse of Hunger with an Eco-Mind." *Journal of Peasant Studies* 40, no. 1 (2013): 219–38.

Lappé, Frances Moore, Joseph Collins, and Cary Fowler. *Food First: Beyond the Myth of Scarcity*. New York: Ballantine Books, 1984.

Larder, Nicolette, Sarah Sippel, and Geoffrey Lawrence. "Finance Capital, Food Security Narratives and Australian Agricultural Land." *Journal of Agrarian Change* 15, no. 4 (2015): 592–603.

LaVecchia, Olivia. "These Neighbors Got Together to Buy Vacant Buildings. Now They're Renting to Bakers and Brewers." *YES Magazine*, September 4, 2015.

http://www.yesmagazine.org/new-economy/neighbors-got-together-buy-vacant-buildings-renting-bike-shop-brewer.

Lazonick, William, and Mary O'Sullivan. "Maximizing Shareholder Value: A New Ideology for Corporate Governance." *Economy and Society* 29, no. 1 (2000): 13–35.

Leivestad, Høyer, and Anette Nyqvist, eds. *Ethnographies of Conferences and Trade Fairs: Shaping Industries, Creating Professionals*. London: Palgrave Macmillan, 2017.

Lerrer, Debora, and John Wilkinson. "Impact of Restrictive Legislation and Popular Opposition Movements on Foreign Land Investments in Brazil: The Case of the Forestry and Pulp Paper Sector and Stora Enso." Paper presented at the International Conference on Global Land Grabbing II, Cornell University, Ithaca, NY, October 2012.

Levien, Michael. *Dispossession without Development: Land Grabs in Neoliberal India*. Oxford: Oxford University Press, 2018.

Lewontin, R. C., and Jean-Pierre Berlan. "Technology, Research, and the Penetration of Capital: The Case of U.S. Agriculture." *Monthly Review* 38, no. 3 (1986): 21–34.

Leyshon, Andrew, and Nigel Thrift. "The Capitalization of Almost Everything: The Future of Finance and Capitalism." *Theory, Culture and Society* 24, no. 7 (2007): 97–115.

Li, Tania. "Centering Labor in the Land Grab Debate." *Journal of Peasant Studies* 38, no. 2 (2011): 281–98.

——. "Rendering Land Investible: Five Notes on Time." *Geoforum* 82 (2017): 276–78.

——. "Transnational Farmland Investment: A Risky Business." *Journal of Agrarian Change* 15, no. 4 (2015): 560–68.

——. "What Is Land? Assembling a Resource for Global Investment." *Transactions of the Institute of British Geographers* 39, no. 4 (2014): 589–602.

Lima, Ruy Cirne. *Pequena história territorial do Brasil: Sesmarias e terras devolutas*. 2nd ed. Porto Alegre, Brazil: Livraria Sulina, 1954.

Lindert, Peter H. "Long-Run Trends in American Farmland Values." *Agricultural History* 62, no. 3 (1988): 45–85.

Lindoso, Felipe. "CPI comprova irregularidades." *O Estado de São Paulo*, May 1, 1968.

LiPuma, Edward, and Benjamin Lee. *Financial Derivatives and the Globalization of Risk*. Durham, NC: Duke University Press, 2004.

Locke, John. *Second Treatise of Government*. Indianapolis: Hackett, 1980. First published in 1690. https://www.gutenberg.org/files/7370/7370-h/7370-h.htm.

Long, Heather. "Nearly Half of America Doesn't Benefit from Dow 22,000." *Washington Post*, August 2, 2017. https://www.washingtonpost.com/news/wonk/wp/2017/08/02/nearly-half-of-america-doesnt-benefit-from-dow-22000/?utm_term=.ed3d60b2e7c6.

*Los Angeles Times*. "New Farming Job Created." March 21, 1926.

MacKenzie, Donald. *An Engine, Not a Camera: How Financial Models Shape Markets*. Cambridge, MA: MIT Press, 2006.

Mackintosh, James, and Kate Burgess. "Hedge Funds Muck In Down on the Farm." *Financial Times*, August 25, 2008. https://www.ft.com/content/94a1ef16-12ff-11dd-8d91-0000779fd2ac.

Magdoff, Harry, and Paul Sweezy. *Stagnation and the Financial Explosion*. New York: Monthly Review, 2009.

Magnan, André. "The Financialization of Agri-food in Canada and Australia: Corporate Farmland and Farm Ownership in the Grains and Oilseed Sector." *Journal of Rural Studies* 41 (2015): 1–12.

Magnan, André, and Sean Sunley. "Farmland Investment and Financialization in Saskatchewan, 2003–2014: An Empirical Analysis of Farmland Transactions." *Journal of Rural Studies* 49 (2017): 92–103.

Malkin, Elisabeth. "Thousands in Mexico City Protest Rising Food Prices." *New York Times*, February 1, 2007. http://www.nytimes.com/2007/02/01/world/americas/01mexico.html.

Malthus, Thomas. *An Essay on the Principle of Population*. London: J. Johnson, 1798. http://www.econlib.org/library/Malthus/malPopCover.html.

Mann, Susan. *Agrarian Capitalism in Theory and Practice*. Chapel Hill: University of North Carolina Press, 1990.

Mann, Susan, and James Dickinson. "Obstacles to the Development of a Capitalist Agriculture." *Journal of Peasant Studies* 5, no. 4 (1978): 466–81.

Marcus, George. "Ethnography in/of the World System: The Emergence of Multi-Sited Ethnography." *Annual Review of Anthropology* 24 (1995): 95–117.

Margolin, Madison. "This Group Wants to Help Lower Rents by Buying Real Estate as a Collective." *Village Voice*, October 17, 2015. https://www.villagevoice.com/2015/10/07/this-group-wants-to-help-lower-rents-by-buying-real-estate-as-a-collective/.

Martin, Randy. *Financialization of Daily Life*. Philadelphia: Temple University Press, 2002.

Martin, Sarah, and Jennifer Clapp. "Finance for Agriculture or Agriculture for Finance?" *Journal of Agrarian Change* 25, no. 4 (2015): 549–59.

Marx, Karl. *Capital: A Critique of Political Economy*. Vol. 1. Edited by Frederick Engels and Ernest Untermann; translated by Samuel Moore and Edward Aveling. Chicago: Charles H. Kerr, 1906. http://www.econlib.org/library/YPDBooks/Marx/mrxCpACover.html.

——. *Capital: A Critique of Political Economy*. Vol. 3. Edited by Frederick Engels; translated by Ernest Untermann. Chicago: Charles H. Kerr, 1909. http://www.econlib.org/library/YPDBooks/Marx/mrxCpCCover.html.

——. *Theories of Surplus Value*. Moscow: Progress, 1963. https://www.marxists.org/archive/marx/works/1863/theories-surplus-value/.

Mason, James R., Jr. "'PSSST, Hey Buddy, Wanna Buy a Country?': An Economic and Political Policy Analysis of Federal and State Laws Governing Foreign Ownership of United States Real Estate." *Vanderbilt Journal of Transnational Law* 27 (1994): 453–88.

Massey, Doreen. "The Pattern of Landownership and Its Implications for Policy." *Built Environment* 6, no. 4 (1980): 263–71.

Massey, Doreen, and Alejandrina Catalano. *Capital and Land: Landownership by Capital in Great Britain*. London: Edward Arnold, 1978.

Masters, Michael. "Testimony of Michael W. Masters before the Committee on Homeland Security and Governmental Affairs, United States Senate." Committee on Homeland Security and Governmental Affairs, May 20, 2008. http://www.hsgac.senate.gov//imo/media/doc/052008Masters.pdf?attempt=2.

Masters, Michael, and Adam White. "The Accidental Hunt Brothers: How Institutional Investors Are Driving Up Food and Energy Prices." Masters Capital Management and White Knight Research and Trading, July 31, 2008. http://www.loe.org/images/content/080919/Act1.pdf.

Mazzucato, Mariana. *The Value of Everything: Making and Taking in the Global Economy*. London: Allen Lane, 2018.

MB Agro and Agroconsult. "Impactos econômicos do parecer da AGU (Advocacia Geral da União), que impõe restrições à aquisição e arrendamento de terras agrícolas por empresas brasileiras com controle do capital detido por estrangeiros." São Paulo: Associação Braileira de Marketing Rural & Agronegócio (ABMR&A), April 2011.

McCarthy, John. "Processes of Inclusion and Adverse Incorporation: Oil Palm and Agrarian Change in Sumatra, Indonesia." *Journal of Peasant Studies* 37, no. 4 (2010): 821–50.

McConnell, Tristan. "The Secret Sale of Southern Sudan." *London Times*, July 2, 2011. http://www.thetimes.co.uk/tto/news/world/africa/article3081861.ece.

McDonald, Michael, and Tatiana Freitas. "Harvard's Foreign Farmland Investment Mess." *Bloomberg Businessweek*, September 6, 2018. https://www.bloomberg.com/news/articles/2018-09-06/harvard-s-foreign-farmland-investment-mess.

McFarlane, Sarah. "Pension Funds to Bulk Up Farmland Investments." Reuters, June 29, 2010. http://uk.reuters.com/article/2010/06/29/uk-pensions-farmland-idUKLNE65S01K20100629.

McGrath, Charles. "Majority of Public Plans Assume Lower Returns in 2018." *Pensions and Investments*, April 27, 2018. https://www.pionline.com/article/20180427/INTERACTIVE/180429882/majority-of-public-plans-assume-lower-returns-in-2018

——. "New York, the Hedge Fund Capital of the Universe." *Pensions and Investments*, February 15, 2018. https://www.pionline.com/article/20180215/INTERACTIVE/180219915/new-york-the-hedge-fund-capital-of-the-universe.

McIntosh, Bill. "Aquila Capital: Absolute Return and Real Asset Strategies from Germany." *Hedge Fund Journal*, December 17, 2010. http://www.thehedgefundjournal.com/node/7524.

McMichael, Philip. "Agrofuels in the Food Regime." *Journal of Peasant Studies* 37, no. 4 (2010): 609–29.

——. "Land Grabbing as Security Mercantilism in International Relations." *Globalizations* 10, no. 1 (2013): 47–64.

Mehta, Lyla. "The Scare, Naturalization and Politicization of Scarcity." In *The Limits to Scarcity: Contesting the Politics of Allocation*, edited by Lyla Mehta, 13–30. London: Earthscan, 2010.

Melichar, Emanuel. *Agricultural Finance Databook*. Washington, DC: Division of Research and Statistics, Board of Governors of the Federal Reserve System, 1987.

Mendonça, Maria Luisa, and Fábio Pitta. "International Financial Capital and the Brazilian Land Market." *Latin American Perspectives* 45, no. 5 (2018): 88–101.

Menegaz, Felipe. "Brazil Labelled Map." Wikimedia Commons, June 11, 2007. https://commons.wikimedia.org/wiki/File:Brazil_Labelled_Map.svg.

Merenlender, Adina, Lynn Huntsinger, Greig Guthey, and S. K. Fairfax. "Land Trusts and Conservation Easements: Who Is Conserving What for Whom?" *Conservation Biology* 18, no. 1 (2004): 65–76.

Meyer, Gregory. "The Great Land Rush: Investors Face Conflict in Quest for Farms." *Financial Times*, March 2, 2016. https://www.ft.com/content/84a646a0-dedc-11e5-b67f-a61732c1d025.

Mezzadra, Sandro, and Brett Neilson. "Operations of Capital." *South Atlantic Quarterly* 114, no. 1 (2015): 1–9.

Mill, John Stuart. *Principles of Political Economy with Some of Their Applications to Social Philosophy*. 7th ed. London: Longmans, Green, 1909. First published in 1848. http://www.econlib.org/library/Mill/mlPCover.html.

Minaya, Jose, and Justin Ourso. "U.S. Drought Shouldn't Scorch Long-Term Farmland Investing." TIAA-CREF, August 14, 2012. https://www.tiaa.org/public/assetmanagement/insights/commentary-perspectives/perspectives/mc_053.html.

Mollo, Maria de Lourdes Rollemberg, and Alfredo Saad-Filho. "Neoliberal Economic Policies in Brazil (1994–2005): Cardoso, Lula and the Need for a Democratic Alternative." *New Political Economy* 11, no. 1 (2006): 99–123. doi:10.1080/13563460500494933.

Mooney, Patrick. "Labor Time, Production Time and Capitalist Development in Agriculture: A Reconsideration of the Mann-Dickinson Thesis." *Sociologia Ruralis* 22, no. 3–4 (1982): 279–92.

———. *My Own Boss? Class, Rationality, and the Family Farm.* Boulder, CO: Westview, 1988.

Morris, Samuel. "Analysis: Global Farmland in a Portfolio Context." *Food for Thought: Macquarie Agricultural Funds Management Newsletter*, December 2012, 8–11.

Morrison, Fred. "Limitations on Alien Investment in American Real Estate." *Minnesota Law Review* 60 (1975): 621–68.

Morrison, Fred, and Kenneth Krause. *State and Federal Legal Regulation of Alien and Corporate Land Ownership and Farm Operation.* Washington, DC: United States Department of Agriculture and Economic Research Service, 1975.

Multilateral Investment Guarantee Agency (MIGA). "MIGA and Chayton Capital LLP to Support Agribusiness Investments in Southern Asia." Press release, May 10, 2010. https://www.miga.org/Lists/General/CustomDisp.aspx?ID=985.

———. "MIGA Reinsures Agribusiness Investment in Zambia." Press release, May 22, 2017. https://www.miga.org/Lists/Press%20Releases/CustomDisp. aspx?ID=533.

Murphy, Sophia, David Burch, and Jennifer Clapp. "Cereal Secrets." Oxfam, 2012.

Nader, Laura. "Up the Anthropologist: Perspectives Gained from Studying Up." In *Reinventing Anthropology*, edited by Dell H. Hyms, 284–311. New York: Pantheon Books, 1972.

Nair, P. K. Ramachandran, Vimala Nair, B. Mohan Kumar, and Julia Showalter. "Carbon Sequestration in Agroforestry Systems." *Advances in Agronomy*, 108, no. 1 (2010): 237–307.

Nasdaq. "Historical Prices." http://www.nasdaq.com/quotes/historical-quotes.aspx.

National Agricultural Law Center. "Corporate Farming Laws—an Overview." http://nationalaglawcenter.org/overview/corporatefarminglaws/.

National Council of Real Estate Investment Fiduciaries (NCREIF). "Farmland Property Index." https://www.ncreif.org/.

National Mining Association. "Historical Gold Prices: 1833 to Present." 2016. https://nma.org/wp-content/uploads/2016/09/historic_gold_prices_1833_pres.pdf.

Native Land Digital. "Native Land." https://native-land.ca/.

Nickerson, Cynthia, Mitchell Morehart, Todd Kuethe, Jayson Beckman, Jennifer Ifft, and Ryan Williams. "Trends in U.S. Farmland Values and Ownership." Washington, DC: United States Department of Agriculture and Economic Research Service, 2012.

Nolte, Kerstin, Wytske Chamberlain, and Markus Giger. *International Land Deals for Agriculture. Fresh Insights from the Land Matrix: Analytical Report II.* Bern: Bern Open, 2016. https://landmatrix.org/media/filer_public/ab/c8/abc8b563-9d74-4a47-9548-cb59e4809b4e/land_matrix_2016_analytical_report_draft_ii.pdf.

Nuveen. "By the Numbers." https://www.nuveen.com/en-us/about-nuveen/nuveen-by-the-numbers.

———. *Responsible Investment in Farmland.* TIAA, 2017. https://www.tiaa.org/public/pdf/06-2017_GBR-CFARMRPT_Farmland_Report.pdf.

———. *Responsible Investment in Farmland.* TIAA, 2018. https://documents.nuveen.com/Documents/Institutional/Default.aspx?fileId=73993.

———. "Westchester Group Investment Management." https://www.nuveen.com/institu-tional/investment-specialists/westchester-group-investment-management.

Oakland Institute. *Down on the Farm.* Edited by Anuradha Mittal and Melissa Moore. Oakland, CA: Oakland Institute, 2014. https://www.oaklandinstitute.org/sites/oaklandinstitute.org/files/OI_Report_Down_on_the_Farm.pdf.

Odilla, Fernanda. "Estrangeiros compram 22 campos de futebol por hora." *Folha de São Paulo*, November 2, 2010. http://www1.folha.uol.com.br/fsp/poder/po0211201002.htm.

Ogliari, Elder. "Compra de terras por múlti no RS reabre debate sobre fronteiras." *O Estado de São Paulo*, March 4, 2008, A10.

O'Keefe, Brian. "Betting the Farm." *Fortune*, June 16, 2009. http://archive.fortune.com/2009/06/08/retirement/betting_the_farm.fortune/index.htm.

Oliveira, Ariovaldo Umbelino. "A questão da aquisição de terras por estrangeiros no Brasil: Um retorno aos dossiês." *Agrária*, no. 12 (2010): 3–113.

Oliveira, Gustavo de L.T. "Chinese Land Grabs in Brazil? Sinophobia and Foreign Investments in Brazilian Soybean Agribusiness." *Globalizations* 15, no. 1(2017): 114–33.

———. "Land Regularization in Brazil and the Global Land Grab." *Development and Change* 44, no. 2 (2013): 261–83.

Oliver, Daniel, Julianne Serovich, and Tina Mason. "Constraints and Opportunities with Interview Transcription: Towards Reflection in Qualitative Research." *Social Forces* 84, no. 2 (2005): 1273–89.

*180 Graus.* "Preço da terra na região da soja no PI passou de R$ 5.800 para R$ 6.100." January 21, 2014. https://180graus.com/aquiles-nairo/preco-da-terra-na-regiao-da-soja-do-piaui-passou-de-r-5800-para-r-6100.

Orange County Employees Retirement System (OCERS). "Investment Manager Moni-toring Subcommittee Meeting." Meeting minutes, Orange County Employees Retirement System, September 24, 2015. https://www.ocers.org/board-committee-meetings.

Organisation for Economic Co-operation and Development (OECD). "Pension Funds in Figures." June 2018. https://www.oecd.org/daf/fin/private-pensions/Pension-Funds-in-Figures-2018.pdf.Ofstehage, Andrew. "Farming Is Easy, Becoming Bra-zilian Is Hard: North American Soy Farmers' Social Values of Production, Work and Land in Soylandia." *Journal of Peasant Studies* 43, no. 2 (2016): 442–60.

Orhangazi, Özgür. *Financialization and the U.S. Economy.* Cheltenham, UK: Edward Elgar, 2008.

Orr, Richard. "Pension-Fund Farm Ventures Are Proposed." *Chicago Tribune*, May 18, 1980.

Osório Silva, Lígia. *Terras devolutas e latifúndio: Efeitos da lei de 1850.* Campinas, Brazil: Editora da Unicamp, 1996.

Ouma, Stefan. "From Financialization to Operations of Capital: Historicizing and Dis-entangling the Finance-Farmland-Nexus." *Geoforum* 72 (2016): 82–93.

———. "Getting in between M and M' or: How Farmland Further Debunks Financial-ization." *Dialogues in Human Geography* 5, no. 2 (2015): 225–28.

———. "Situating Global Finance in the Land Rush Debate: A Critical Review." *Geofo-rum* 57 (2014): 162–66.

———. "This Can('t) Be an Asset Class: The World of Money Management, 'Society,' and the Contested Morality of Farmland Investments." *Environment and Plan-ning A.* OnlineFirst. DOI: 10.1177/0308518X18790051.

Paasi, Anssi. "Boundaries as Social Processes: Territoriality in the World of Flows." *Geopolitics* 3, no. 1 (1998): 69–88. doi:10.1080/14650049808407608.

Panini, Carmela. *Reforma agrária dentro e fora da lei: 500 anos de historia inacabada.* São Paulo: Edições Paulinas, 1990.

Payne, Susan. "The Case for African Agriculture: Asymmetries Dispelled, with Case Studies." Speech delivered at 2013 Land Investment Expo, West Des Moines, IA, January 18, 2013. https://youtu.be/9BoCwc9WOs8.

Penniman, Leah. *Farming While Black: Soul Fire Farm's Practical Guide to Liberation on the Land.* White River Junction, VT: Chelsea Green, 2018.

Peoples, Kenneth, David Freshwater, Gregory Hanson, Paul Prentice, and Eric Thor. *Anatomy of an American Agricultural Credit Crisis: Farm Debt in the 1980s.* Lanham, MD: Rowman & Littlefield, 1992.

Pereira, Rosalinda P. C. Rodrigues. "A teoria da função social da propriedade rural e seus reflexos na acepção clássica de propriedade." In *A questão agrária e a justiça,* edited by Juvelino José Strozake, 88–129. São Paulo: Editora Revista dos Tribunais, 2000.

Peres, Eraldo, and Sarah Dilorenzo. "Brazil's Indigenous Protest to Defend Their Rights, Lands." *Washington Post,* April 26, 2018.

Phelps, William, Jr. "Corporate Farming Statutes." *Whittier Law Review* 2 (1979): 441–67.

Pike, Andy, and Jane Pollard. "Economic Geographies of Financialization." *Economic Geography* 86, no. 1: 29–51.

Piketty, Thomas. *Capital in the Twenty-First Century.* Cambridge, MA: Harvard University Press, 2014.

Pini, Barbara. "Interviewing Men: Gender and the Collection and Interpretation of Qualitative Data." *Journal of Sociology* 41, no. 2 (2005): 201–16.

Plata, Ludwig. "Mercados de terras no Brasil: Gênese, determinação de seus preços e políticas." PhD diss., Universidade Estadual de Campinas (UNICAMP), 2001.

Polanyi, Karl. *The Great Transformation: The Political and Economic Origins of Our Time.* Boston: Beacon, 2001.

Pollard, Jane, Jonathan Oldfield, Samuel Randalls, and John Thornes. "Firm Finances, Weather Derivatives and Geography." *Geoforum* 39 (2008): 616–24.

Popper, Nathaniel. "Restyled as Real Estate Trusts, Varied Businesses Avoid Taxes." *New York Times,* April 21, 2013. https://www.nytimes.com/2013/04/22/business/restyled-as-real-estate-trusts-varied-businesses-avoid-taxes.html.

Potter, Paul. "Farm Managers Find Means to Boost Income." *Chicago Tribune,* August 28, 1932.

Povinelli, Elizabeth. *The Cunning of Recognition: Indigenous Alterities and the Making of Australian Multiculturalism.* Durham, NC: Duke University Press, 2002.

Prager, Alicia, and Flávia Milhorance. "Cerrado: U.S. Investment Spurs Land Theft, Deforestation in Brazil, Say Experts." *Mongabay,* March 28, 2018. https://news.mongabay.com/2018/03/cerrado-u-s-investment-spurs-land-theft-deforestation-in-brazil-say-experts/.

Preqin. "Preqin Global Natural Resources Report Data Pack." 2017.

———. "Preqin Special Report: Agriculture Data Pack." September 2016.

Pretto, José Miguel. "Imóveis rurais sob propriedade de estrangeiros no Brasil." Brasília: Núcleo de Estudos Agrários e Desenvolvimento Rural (NEAD), 2009.

Principles for Responsible Investment (PRI). *Responsible Investment in Farmland: Report 2014–2015.* 2015. https://www.unpri.org/download_report/6243.

Prudential Global Investment Management. "Agricultural Equity." http://www.pgimref.com/real-estate-finance/businesses-agricultural-equity-investments.shtml.

Radin, Margaret Jane. *Contested Commodities: The Trouble with Trade in Sex, Children, Body Parts, and Other Things.* Cambridge, MA: Harvard University Press, 1996.

Randalls, Samuel. "Weather Profits: Weather Derivatives and the Commercialization of Meteorology." *Social Studies of Science* 40, no. 5 (2010): 705–30.

Rede Social de Justiça e Direitos Humanos, GRAIN, Inter Pares, and Solidarity Sweden. "Foreign Pension Funds and Land Grabbing in Brazil." November 16, 2015. https://www.grain.org/article/entries/5336-foreign-pension-funds-and-land-grabbing-in-brazil.

Reis, Elisa P. "Brazil: One Hundred Years of the Agrarian Question." *International Social Science Journal* 50, no. 157 (1998): 419–32.

Rescher, Nicholas. *Luck: The Brilliant Randomness of Everyday Life*. Pittsburgh: University of Pittsburgh Press, 1995.

Reydon, Bastiaan. "A regulação institucional da propriedade da terra no Brasil: Uma necessidade urgente." In *Dimensões do agronegócio brasileiro políticas, instituições e perspetivas*, edited by Pedro Ramos, 226–62. Brasília: Ministério do Desenvolvimento Agrário (MDA), 2007.

——. Introduction to *Mercados de terras no Brasil: Estrutura e dinâmica*, edited by Bastiaan Reydon and Francisca Cornélio, 15–21. Brasília: Ministério do Desenvolvimento Agrário (MDA) and Núcleo de Estudos Agrários e Desenvolvimento Rural (NEAD), 2006.

Reydon, Bastiaan, Vitor Fernandes, and Tiago Telles. "Land Tenure in Brazil: The Question of Regulation and Governance." *Land Use Policy* 42 (2015): 509–16.

Reydon, Bastiaan, and Ludwig Plata. *Intervenção estatal no mercado de terras: A experiência recente no Brasil*. Campinas, Brazil: UNICAMP, 2000.

Reydon, Bastiaan, Ademar Romeiro, Ludwig Plata, and Marcos Soares. "Preço elevado e o ITR." In *Mercados de terras no Brasil: Estrutura e dinâmica*, edited by Bastiaan Reydon and Francisca Cornélio, 155–78. Brasília: Ministério do Desenvolvimento Agrário (MDA) and Núcleo de Estudos Agrários e Desenvolvimento Rural (NEAD), 2006.

Reydon, Bastiaan, Edar da Silva Anãnã, Gilberto de Oliveira Kloeckner, and Francisca Cornélio. "Ativo terra agrícola em carteiras de investimento." In *Mercados de terras no Brasil: Estrutura e dinâmica*, edited by Bastiaan Reydon and Francisca Cornélio, 181–204. Brasília: Ministério do Desenvolvimento Agrário (MDA) and Núcleo de Estudos Agrários e Desenvolvimento Rural (NEAD), 2006.

Rezende, Gervásio Castro de. "Crédito rural subsidiado e preço da terra no Brasil." *Estudos Econômicos* 12, no. 2 (1982): 117–37.

Ricardo, David. *On the Principles of Political Economy and Taxation*. 3rd ed. London: John Murray, 1821. First published in 1817. http://www.econlib.org/library/Ricardo/ricPCover.html.

Roche, Maurice. "Fads versus Fundamentals in Farmland Prices: Comment." *American Journal of Agricultural Economics* 83, no. 4 (2001): 1074–77.

Rochedo, Pedro, Britaldo Soares-Filho, Roberto Schaeffer, Eduardo Viola, Alexandre Szklo, André Lucena, Alexandre Koberle, Juliana Davis, Raoni Rajão, and Regis Rathmann. "The Threat of Political Bargaining to Climate Mitigation in Brazil." *Nature Climate Change* 8 (2018): 695–98.

Rogers, Barclay. "How Agtech Has Quietly Transformed Agriculture during the Downturn." *AgFunder News*, August 27, 2018. https://agfundernews.com/agtech-quietly-transformed-agriculture-downturn.html.

Romero, Simon. "TIAA-CREF, U.S. Investment Giant, Accused of Land Grabs in Brazil." *New York Times*, November 16, 2015. https://www.nytimes.com/2015/11/17/world/americas/tiaa-cref-us-investment-giant-accused-of-land-grabs-in-brazil.html?_r=0.

Rosenberg, Greg, and Jeffrey Yuen. "Beyond Housing: Urban Agriculture and Commercial Development by Community Land Trusts." Lincoln Institute of Land Policy Working Paper. Cambridge, MA: Lincoln Institute of Land Policy, 2012.

Ross, Andrew. "The Lonely Hour of Scarcity." *Capitalism Nature Socialism* 7, no. 3 (1996): 3–26.

Rota Fortunae. "Farmland Partners: Loans to Related-Party Tenants Introduce Significant Risk of Insolvency—Shares Uninvestible." Seeking Alpha, July 11, 2018. https://seekingalpha.com/article/4186457-farmland-partners-loans-related-party-tenants-introduce-significant-risk-insolvency-shares.

Rotz, Sarah, Emily Duncan, Matthew Small, Janos Botschner, Rozita Dara, Ian Mosby, Mark Reed, and Evan Fraser. "The Politics of Digital Agricultural Technologies: A Preliminary Review." *Sociologia Ruralis* 59, no. 2 (2019): 203–29.

Rural Funds Group. "Assets." https://ruralfunds.com.au/rural-funds-group/fund-information/assets/.

——. "RFF Fund Overview." https://ruralfunds.com.au/rural-funds-group/fund-information/fund-overview/.

Russi, Luigi. *Hungry Capital: The Financialization of Food*. Winchester, UK: Zero Books, 2013.

Said, Edward. *Orientalism*. New York: Pantheon Books, 1978.

Saloutos, Theodore, and John Donald Hicks. *Agricultural Discontent in the Middle West, 1900–1939*. Charleston, NC: Nabu, 1951.

Sant'Anna, Lourival. "Chineses desistem de plantar e agora financiam e exportam soja brasileira." *O Estado de São Paulo*, January 2, 2014. http://economia.estadao.com.br/noticias/geral,chineses-desistem-de-plantar-e-agora-financiam-e-exportam-soja-brasileira-imp-,1114200.

Sauer, Sérgio, and Sérgio Pereira Leite. "Agrarian Structure, Foreign Investment in Land, and Land Prices in Brazil." *Journal of Peasant Studies* 39, no. 3 (2012): 873–98.

Sautchuk, Jaime, Horácio Martins de Carvalho, and Sérgio Buarque de Gusmão. *Projeto Jari: A invasão americana*. São Paulo: Brasil Debates, 1979.

Savills Research. "Global Farmland Index." June 2016. http://pdf.euro.savills.co.uk/uk/rural---other/global-farmland-index.pdf.

——. "International Farmland Focus 2012." 2012. http://pdf.euro.savills.co.uk/global-research/international-farmland-focus.pdf.

——. "International Farmland Focus 2014." 2014. https://pdf.euro.savills.co.uk/rural-other/int-farmland-lores.pdf.

Sayad, João. "Preço da terra e mercados financeiros." *Pesquisa e Planejamento Econômico* 7, no. 3 (1977): 623–62.

Schmelzkopf, Karen. "Incommensurability, Land Use, and the Right to Space: Community Gardens in New York City." *Urban Geography* 23, no. 4 (2002): 323–43.

Schneider, Keith. "As More Family Farms Fail, Hired Managers Take Charge." *New York Times*, March 17, 1986.

Schnitky, Gary, and Bruce Sherrick. "Income and Capitalization Rate Risk in Agricultural Real Estate Markets." *Choices* 26, no. 2 (2011). http://choicesmagazine.org/choices-magazine/theme-articles/farmland-values/income-and-capitalization-rate-risk-in-agricultural-real-estate-markets.

Schutz, Anthony B. "Corporate-Farming Measures in a Post-Jones World." *Drake Journal of Agricultural Law* 14 (2009): 97–147.

Scolese, Eduardo. "Por dia, estrangeiro compra '6 Mônacos' de terra no país." *Folha de São Paulo*, July 7, 2008. http://www1.folha.uol.com.br/fsp/brasil/fc0707200813.htm.

Scott, James C. *The Moral Economy of the Peasant: Rebellion and Subsistence in Southeast Asia.* New Haven, CT: Yale University Press, 1976.

——. *Seeing Like a State: How Certain Schemes to Improve the Human Condition Have Failed.* New Haven, CT: Yale University Press, 1998.

Sen, Amartya. *Poverty and Famines: An Essay on Entitlement and Deprivation.* Oxford: Oxford University Press, 1982.

Senf, Dave. "Life Insurance Company Ownership of United States' Agricultural Land in 1987." Minneapolis, MN: Center for Urban and Regional Affairs, 1988.

Shideler, James. *Farm Crisis: 1919–1923.* Berkeley: University of California Press, 1957.

Shiller, Robert. "Understanding Recent Trends in House Prices and Home Ownership." Cowles Foundation Discussion Paper. New Haven, CT: Cowles Foundation for Research in Economics, September 2007. http://dl4a.org/uploads/pdf/d1630.pdf.

Shipton, Parker. *Mortgaging the Ancestors: Ideologies of Attachment in Africa.* New Haven, CT: Yale University Press, 2009.

Sippel, Sarah. "Financialising Farming as a Moral Imperative? Renegotiating the Legitimacy of Land Investments in Australia." *Environment and Planning A* 50, no. 3 (2018): 549–68.

——. "Food Security or Commercial Business? Gulf State Investments in Australian Agriculture." *Journal of Peasant Studies* 42, no. 5 (2015): 981–1001.

Sippel, Sarah, Nicolette Larder, and Geoffrey Lawrence. "Grounding the Financialization of Farmland: Perspectives on Financial Actors as New Land Owners in Rural Australia." *Agriculture and Human Values* 34, no. 2 (2017): 251–65.

Skidmore, Thomas. *Brazil: Five Centuries of Change.* 2nd ed. Oxford: Oxford University Press, 2010.

SLC Agrícola. "December 2012 Presentation for Investors." December 2012. http://ri.slcagricola.com.br/enu/646/slcagricola_presentation_december_en.pdf.

——. "March 2016 Presentation for Investors." March 2016. http://ri.slcagricola.com.br/enu/1839/SLC_Apresentacao%20RI_PT_BOOK_Mar_2016_eng.pdf.

Smith, Adam. *An Inquiry into the Nature and Causes of the Wealth of Nations.* 5th ed. Edited by Edwin Cannan. London: Methuen and Co., 1904. First published in 1776. http://www.econlib.org/library/Smith/smWN5.html#B.I.

Sommerville, Melanie, and André Magnan. "'Pinstripes on the Prairies': Examining the Financialization of Farming Systems in the Canadian Prairie Provinces." *Journal of Peasant Studies* 42, no. 1 (2015).

South of the Sound Community Farm Land Trust. "Community Farm Land Trust." https://www.communityfarmlandtrust.org/.

Stam, Jerome, Steven Koenig, and George Wallace. *Life Insurance Company Mortgage Lending to US Agriculture: Challenges and Opportunities.* Washington, DC: US Department of Agriculture and Economic Research Service, 1995.

Stark, David. "For a Sociology of Worth." Center on Organizational Innovation, October 2000.

Stephens, Phoebe. "The Principles of Responsible Agricultural Investment." *Globalizations* 10, no. 1 (2013): 187–92.

Stiglitz, Joseph. "Capital Market Liberalization, Globalization, and the IMF." In *Capital Market Liberalization and Development,* edited by José Ocampo and Joseph Stiglitz, 76–100. Oxford: Oxford University Press, 2008.

——. *The Roaring Nineties.* New York: W. W. Norton, 2003.

Stooker, Richard. *REITs around the World.* Gold Egg Investing, 2013.

Stookey, Hunt, and Philippe de Lapérouse. 2009. "Agricultural Land Investment." HighQuest Partners. http://www.highquestpartners.com/userfiles/files/AgLand_Investing.pdf.

Stora Enso. "Shareholders and Ownership Changes." http://www.storaenso.com/investors/shares/shareholders-and-ownership-changes.

Sustainable Iowa Land Trust. "SILT." https://silt.org/.

Tan, Richard Ming Kirk. "Restrictions on the Foreign Ownership of Property: Indonesia and Singapore Compared." *Journal of Property Investment & Finance* 22, no. 1 (2004): 101–11.

Taylor, Jerome. "Fields of Gold: Investors Discover Lucrative Haven in Britain's Farmland." *Independent*, April 17, 2008. https://www.independent.co.uk/news/uk/this-britain/fields-of-gold-investors-discover-lucrative-haven-in-britains-farmland-810376.html.

Taylor, Linnet. "Data Subjects or Data Citizens? Addressing the Global Regulatory Challenge of Big Data." In *Information, Freedom and Property: The Philosophy of Law Meets the Philosophy of Technology*, edited by Mireille Hildebrandt and Bibi van den Berg, 81–104. New York: Routledge, 2016.

Tharpar, Neil. "The Future of Farmland (Part 2): Grabbing the Land Back." Sustainable Economies Law Center. June 22, 2017. https://www.theselc.org/reit_blog_part_2.

Thompson, E. P. *The Making of the English Working Class*. New York: Vintage Books, 1963.

TIAA. *Responsible Investment in Farmland*. 2015. https://www.tiaa.org/public/pdf/C26304_2015_Farmland_Report.pdf.

——. *Responsible Investment in Farmland*. 2016. https://www.tiaa.org/public/pdf/2016_Farmland_Report_FINAL_2.pdf.

TIAA-CREF. *Responsible Investment in Farmland*. 2012. http://supplement.pionline.com/real-assets/_pdf/TIAA-CREF_responsible_investment.pdf (site discontinued).

——. "TIAA-CREF Announces $2 Billion Global Agriculture Company." Press release, May 14, 2012. http://www.businesswire.com/news/home/20120514006381/en/TIAA-CREF-Announces-2-Billion-Global-Agriculture-Company.

——. "TIAA-CREF Announces $3 Billion Global Agriculture Investment Partnership." Press release, August 4, 2015. https://www.tiaa.org/public/about-tiaa/news-press/press-releases/pressrelease602.html.

Tillable. "For Farmland Investors." https://tillable.com/for-farmland-investors/.

Tirole, Jean. "Asset Bubbles and Overlapping Generations." *Econometrica* (1985): 1071–1100.

Tomaskovic-Devey, Donald, and Ken-Hou Lin. "Income Dynamics, Economic Rents, and the Financialization of the U.S. Economy." *American Sociological Review* 76, no. 4 (2011): 538–59.

Trevisan, Cláudia. "Estatal da China quer produzir soja no país." *O Estado de São Paulo*, April 22, 2010.

Tsing, Anna. *Friction: An Ethnography of Global Connection*. Princeton, NJ: Princeton University Press, 2005.

Tuckett, David. "Financial Markets Are Markets in Stories: Some Possible Advantages of Using Interviews to Supplement Existing Economic Data Sources." *Journal of Economic Dynamics and Control* 36 (2012): 1077–87.

Turvey, Calum. "Hysteresis and the Value of Farmland: A Real-Options Approach to Farmland Valuation." In *Government Policy and Farmland Markets: The Maintenance of Farmer Wealth*, edited by Charles Moss and Andrew Schmitz, 179–207. Ames: University of Iowa Press, 2003.

US Agriculture. "US Agriculture and Halderman Real Asset Management Announce Merger." Press release, April 25, 2016. http://www.us-agriculture.com/files/press_releases/US_Agriculture_Press_Release_%202016-04-25.pdf.

US Congress. "Ag-Land Trust Proposal: Hearings before the Subcommittee on Family Farms, Rural Development, and Special Studies of the Committee on Agriculture." House of Representatives, Ninety-Fifth Congress, First Session, 1977. https://babel.hathitrust.org/cgi/pt?id=umn.31951d002711810.

——. "The Investment of Pension Funds in Farmland: Hearing before the Select Committee on Small Business." United States Senate, Ninety-Sixth Congress, Second Session on the Investment of Pension Funds in Farmland, October 8, 1980, 1980. https://babel.hathitrust.org/cgi/pt?id=mdp.39015082594162.

US Department of Agriculture Economic Research Service (USDA ERS). "Farm Income and Wealth Statistics." https://www.ers.usda.gov/data-products/farm-income-and-wealth-statistics.aspx.

US Department of Agriculture National Agricultural Statistics Service (USDA NASS). "Quick Stats." https://quickstats.nass.usda.gov/.

US General Accounting Office. "Pension Fund Investment in Agricultural Land," no. 81 (April 2, 1981). http://www.gao.gov/products/114880.

Vaggi, Gianni. "Physiocrats." *The New Palgrave Dictionary of Economics*. 3rd ed., 869–76. New York: Palgrave Macmillan, 1987.

*Valor Econômico*. "MST promete invasões se congresso aprovar venda de terra a estrangeiro." August 6, 2016. http://www.valor.com.br/politica/4661239/mst-promete-invasoes-se-congresso-aprovar-venda-de-terra-estrangeiro.

Vaz, Lúcio. "Aproveitando a flexibilidade da legislação, empresas multinacionais adquirem terras com o consentimento do governo brasileiro." *Correio Braziliense*, June 13, 2010. http://www.correiobraziliense.com.br/app/noticia/brasil/2010/06/13/interna_brasil,197417/index.shtml.

Verdery, Katherine. *The Vanishing Hectare: Property and Value in Postsocialist Transylvania*. Ithaca, NY: Cornell University Press, 2003.

Vieira, André. "Stora Enso obtém aval para regularização de terras no RS." *Valor Econômico*, September 4, 2009.

Visser, Oane. "Running Out of Farmland?" *Agriculture and Human Values* 34 (2017): 185–98.

Visser, Oane, and Max Spoor. "Land Grabbing in Post-Soviet Eurasia: The World's Largest Agricultural Land Reserves at Stake." *Journal of Peasant Studies* 38, no. 2 (2011): 299–323.

Wahl, Peter. "Food Speculation: The Main Factor of the Price Bubble in 2008." Briefing paper. Berlin: World Economy, Ecology, & Development. January 1, 2009. https://www.iatp.org/sites/default/files/451_2_105384.pdf.

Walker, Richard. *The Conquest of Bread: 150 Years of Agribusiness in California*. New York: New Press, 2004.

*Wall Street Journal*. "Chain Farm Method Is Tested in Iowa." November 12, 1930.

Ward, Callum, and Manuel Aalbers. "Virtual Special Issue Editorial Essay: 'The Shitty Rent Business': What's the Point of Land Rent Theory?" *Urban Studies* 53, no. 9 (2016): 1760–83.

Wasik, John. "More Precious Than Gold? Farmland Has Glowing Appeal." Reuters, November 15, 2011. https://www.reuters.com/article/us-usa-investing-farmland-idUSTRE7AE1HQ20111115.

Watts, Michael. "Life under Contract: Contract Farming, Agrarian Restructuring, and Flexible Accumulation." In *Living under Contract: Contract Farming and Agrarian Transformation in Sub-Saharan Africa*, edited by Peter Little and Michael Watts, 21–77. Madison: University of Wisconsin Press, 1994.

——. *Silent Violence: Food, Famine, and Peasantry in Northern Nigeria*. Berkeley: University of California Press, 1983.

Welch, Cliff. "Globalization and the Transformation of Work in Rural Brazil: Agribusiness, Rural Labor Unions, and Peasant Mobilization." *International Labor and Working-Class History* 70, no. 1 (2006): 35–60.

Welsh, Rick, Chantal Carpentier, and Bryan Hubbell. "On the Effectiveness of State Anti-corporate Farming Laws in the United States." *Food Policy* 26 (2001): 543–48.

Whatmore, Sarah. "Landownership Relations and the Development of Modern British Agriculture." In *Agriculture: People and Policies*, edited by Graham Cox, Philip Lowe, and Michael Winter, 105–25. Amsterdam: Springer, 1986.

Wheaton, Bradley, and William Kiernan. "Farmland: An Untapped Asset Class?" *Food for Thought: Macquarie Agricultural Funds Management Newsletter*, December 2012, 3–7.

Wilkinson, John, Bastiaan Reydon, and Alberto Di Sabbato. "Concentration and Foreign Ownership of Land in Brazil in the Context of Global Land Grabbing." *Canadian Journal of Development Studies* 33, no. 4 (2012): 417–38.

——. "Dinâmica no mercado de terras na América Latina: O caso do Brasil." Santiago, Chile: FAO Regional Office for Latin America, 2010.

Williams, James. "Feeding Finance: A Critical Account of the Shifting Relationships between Finance, Food and Farming." *Economy and Society* 43, no. 3 (2014): 401–31.

Willis Towers Watson. "Global Pension Assets Study 2017." January 30, 2017. https://www.willistowerswatson.com/en/insights/2017/01/global-pensions-asset-study-2017.

Wilson, Grant. "Reforming Alien Agricultural Landownership Restrictions in Corporate Farming Law States: A Constitutional and Policy View From Iowa." *Drake Journal of Agricultural Law* 17 (2012): 709–53.

Wilson, Jeff. "Corn Farms Are Hotter Than New York Lofts." *New York Sun*, February 21, 2007. http://www.nysun.com/business/corn-farms-are-hotter-than-new-york-lofts/48972/.

Wily, Liz Alden. "'The Law Is to Blame': The Vulnerable Status of Common Property Rights in Sub-Saharan Africa." *Development and Change* 42, no. 3 (2011): 733–57.

Windsor, Duane. "Corporate Social Responsibility: Three Key Approaches." *Journal of Management Studies* 43, no. 1 (2006): 93–114.

Wise, Murray. *Farmland Investment Strategy: The Opportunity of the 1990s*. Champaign, IL: WGI, 1993.

——. *Investing in Farmland: A Complete Guide to Evaluating, Financing and Managing Income-Producing Farm Properties*. Chicago: Probus, 1989.

Wittman, Hannah. "Reframing Agrarian Citizenship: Land, Life and Power in Brazil." *Journal of Rural Studies* 25, no. 1 (2009): 120–30.

Wittman, Hannah, Jessica Dennis, and Heather Pritchard. "Beyond the Market? New Agrarianism and Cooperative Farmland Access in North America." *Journal of Rural Studies* 53 (2017): 303-16.

Wolf, Eric. *Peasant Wars of the Twentieth Century*. New York: Harper & Row, 1969.

Wolf, Steven and Spencer Wood. "Precision Farming: Environmental Legitimation, Commodification of Information, and Industrial Coordination." *Rural Sociology* 62, no.2 (1997): 180–206.

Wolford, Wendy. "Agrarian Moral Economies and Neoliberalism in Brazil." *Environment and Planning A* 37 (2005): 241–61.

——. *This Land Is Ours Now: Social Mobilization and the Meanings of Land in Brazil*. Durham, NC: Duke University Press, 2010.

Woodruff, Archibald. *Farm Mortgage Loans of Life Insurance Companies.* New Haven, CT: Yale University Press, 1937.

Wray, L. Randall. "The Commodities Market Bubble: Money Manager Capitalism and the Financialization of Commodities." Jerome Levy Economics Institute of Bard College, no. 96 (2008).

Zaia, Cristiano. "Casa Civil quer venda de terra a estrangeiro sem limite de área." *Valor Econômico,* April 6, 2017. http://www.valor.com.br/agro/4928916/casa-civil-quer-venda-de-terra-estrangeiro-sem-limite-de-area.

Zaia, Cristiano, and Fernando Exman. "Ruralistas tentam emplacar venda de terra a estrangeiro." *Valor Econômico,* October 9, 2019. https://valor.globo.com/brasil/noticia/2019/09/10/ruralistas-tentam-emplacar-venda-de-terra-a-estrangeiro.ghtml.

Zaloom, Caitlin. *Out of the Pits: Traders and Technology from Chicago to London.* Chicago: University of Chicago Press, 2006.

Zelizer, Viviana. "Human Values and the Market: The Case of Life Insurance and Death in 19th-Century America." *American Journal of Sociology* 84, no. 3 (November 1978): 591–610.

——. *Pricing the Priceless Child: The Changing Social Value of Children.* Princeton, NJ: Princeton University Press, 1985.

——. *The Purchase of Intimacy.* Princeton, NJ: Princeton University Press, 2005.

Zero Hedge. "Is TIAA-CREF Investing in Farmland a Harbinger of the Next Asset Bubble?" Zerohedge.com, October 4, 2010. http://www.zerohedge.com/article/tiaa-cref-investing-farmland-harbinger-next-asset-bubble.

Zhang, Daowei, Brett J. Butler, and Rao V. Nagubadi. "Institutional Timberland Ownership in the US South: Magnitude, Location, Dynamics, and Management." *Journal of Forestry* 110, no. 7 (2012): 355–61.

Zhang, Wendong. "2016 Farmland Value Survey Iowa State University." Iowa State University Extension and Outreach, December 2016. https://www.extension.iastate.edu/agdm/wholefarm/html/c2-70.html.

Zhou, Nora. "Brookfield Nears First Close of Brazil AgriLand Fund II." Agri Investor, May 18, 2015. https://www.agriinvestor.com/brookfield-nears-first-close-brazil-agriland-fund-ii-exclusive/.

Zoomers, Annelies. "Globalisation and the Foreignisation of Space: Seven Processes Driving the Current Global Land Grab." *Journal of Peasant Studies* 37, no. 2 (2010): 429–47.

Zoomers, Annelies, Femke Van Noorloos, Kei Otsuki, Griet Steel, and Guus Van Westen. "The Rush for Land in an Urbanizing World: From Land Grabbing toward Developing Safe, Resilient, and Sustainable Cities and Landscapes." *World Development* 92 (2017): 242–52.

Zuboff, Shoshana. "Big Other: Surveillance Capitalism and the Prospects of an Information Civilization." *Journal of Information Technology* 30 (2015): 75–89.

# Index

Page numbers in *italics* refer to figures and tables.